Published by:
© 1998 by autovision - verlag
Kronprinzenstr. 54
22587 Hamburg

Authors:
Matthias Pfannmüller, Hamburg
Boris Schmidt, Frankfurt

Translation:
Paul Entwistle, Rover Group Ltd., Warwick

Layout and Production:
Kontor 11 Grafikdesign GmbH, Hamburg
Typeset in FF Scala Sans

Cover Design:
The Drury Lane Company
Heath Farm
Hampton Lane
Meriden
West Midlands
CV7 7LL

Reproduction and Lithography:
Winkelhaken dtp, Hamburg

Printed by:
Mediaprint Informationstechnologie GmbH, Paderborn

Printed in Germany

ISBN 3-9805832-0-1

Contents

Preface

"When the first Land Rover was being put together, we were still little boys. Consequently we feel that we have grown up with it and we are proud of the fact that it was our dad Maurice's idea. We can still picture the way he experimented with the early models, examined every detail for necessary improvements and then notified his team at Rover to change this or that item. Successful products have always come about as a result of the efforts of many individuals and this was no different at Land Rover. From drawing board sketch to production vehicle barely took one year and everyone in the development team had to work very hard to achieve this. Our father was technically very gifted and constantly put his hand to new projects. Once, he built an automatic lawnmower using a car's rear axle and a few bits from an old bike. In the late forties, he made a tape recorder from an old army mine detector. It must have been one of the first tape recorders in domestic use and we had loads of fun with it. The Land Rover too, is a cause for a great deal of childhood nostalgia. At about the time they started up production, our father was laying out a fish pond in the garden. He used a pump fitted to his Land Rover to fill it up with water from a nearby spring. Naturally we boys wanted to help him. However, because the pressure was too high, we could barely hold on to the nozzle on the end of the hose. The result was a stream of water arcing its way into the trees...

Another time I remember we were sitting in the rear load tray of a Land Rover with the kids from next door, when we were stopped by a policeman who was unhappy about the lack of seats in the back and threatened to

Patrick and Steven Wilks with their brother William on the beach at Red Wharf Bay on Anglesey (from left to right)

First hand experience: The brothers Patrick and Steven Wilks (on the right)

book dad. It wasn't long after that all Land Rovers were given optional seating in the back. We can remember numerous other trips out with the off-roader. Although we did not realise at the time, these trips served to improve the vehicle still further. Dad's test vehicles weren't exactly treated lightly either as he liked going to the limit of their capabilities. Once, we were driving down a particularly steep incline when the Land Rover grounded out on a hillock. Dad realised that the fuel tank extended too far below the chassis which prevented the wheels from staying in contact with the ground. For this reason, the tank was later altered resulting in a much flatter design. Another time dad managed to sink his Land Rover while attempting to cross the mud flats going through a tideway. The tide came in and of course it was impossible to recover the vehicle. It wasn't long before the vehicle was under two yards of water and we had to abandon it. Amazingly however we were able to go back the following day, immediately start the engine and recover the vehicle without any difficulty. Every year we went on holiday with a new Land Rover that was better than the previous incarnation. We boys loved the Land Rover, but mum had problems with it at first. She always used to say that the Land Rover was made for men by men. She did not like the sharp edges and the heavy steering. Dad always took these criticisms to heart, not least because mum did a vast number of miles in the Land Rover herself. It did not take long for the vehicle to be made more comfortable. Of course, this was all over fifty years ago, but we haven't forgotten. Today, Land Rovers are developed continuously and it is this which forms the basis of Land Rover's excellent reputation and worldwide popularity. So it is now up to you, the reader, to find out some more about the story of the Land Rover in this book."

Birmingham, August 1997

P. Wilks

S. Wilks.

Traces in the Sand

Anglesey, the largest island off the coast of Great Britain, lies to the north of Colwyn and above the Lleyn peninsula in the north west corner of Wales. The inhabitants of this barren and precipitous isle in the Irish Sea do not like to be called 'Brits'. They are Welsh. Proof positive of this fact alone is their curious tongue, a dialect by which the ancient celts communicated with one another and which is so foreign and unpronounceable to those not versed in it that no one in their right mind would choose to make it their own without having had it rammed down their throats along with their mothers' milk from day one. Who else would comprehend the name of that greatest of railway stations Llanfairpwllgwyngyllgogery-chwyrndrobwllllantsiliogogogoch which is pronounced Llan-vire-pooll-guin-gill-go-ger-u-queern-drob-ooll-llandus-ilio-gogogoch and which is groaned and hissed at breathtaking speed? This tongue twister by the way, tells the magical tale of a valley church situated near a red cave. In keeping with this spirit of linguistic deterrence, there are few towns and street signs which deign to explain themselves in English. It is almost as if Anglesey's population want to keep the island for themselves and deny the existence of foreigners completely. The few that do venture on to the island searching for the ferry port at Holyhead have been known to get hopelessly lost. Such negative first impressions however, would be wildly inaccurate.The locals are hospitable, open-minded and eager to help. They do speak English, well, a sort of English. Conversation really gets into gear in one of the cosy pubs where local music complements beer and fresh fish. Anglesey though, has much more

to offer than this. Megalithic graves from the stone age point to the civilised past of this island known as Mother Wales. In the Middle Ages, the island was a veritable corn belt. Centuries later, precious metals were discovered here. During the industrial revolution, the neighbouring Parys mountain was developed into the world's largest copper mine. Alongside such historical attractions, Anglesey possesses a rich panoply of plant and animal life. Indeed, some species of fauna and flora are unique to the island. In this little world, nature still has its health. The air is clean and smells of salt. Not surprisingly, the islanders are fiercely proud of their unspoilt environment. Clean bays with miles of open beach seemingly designed with the bather in mind, traditionally made Anglesey a popular holiday destination for the English.

It happened one sunny, autumn day in 1947 in Red Wharf Bay near Benllech on the north east corner of the island. At the end of the narrow track leading from the main road to the broad stretch of sand a gaggle of Austins, Vauxhalls, Rileys and of course Rovers, are parked up. All are 1930s vintage characterised by their long bonnets, swept-back, streamlined wings and exposed headlights. It is over two years since the end of World War II. Nevertheless, British industry is still struggling to recover from the war effort. The British automobile industry is also barely capable of producing new models. Lack of capacity and raw materials force manufacturers to draw on pre-war construction techniques. However on this particular afternoon, the ever-present shortages and worries besetting industry seem a long way away. Only the whooping of the children disturbs the peace and quiet. They splash through the shallows and dance between the rays of sunshine. A little way off, the adults are relaxing, eating chicken and mayonnaise sandwiches, reading the Times and every once in a while casting a glance towards the sea.

No one is really aware of what is happening barely one hundred metres down the beach. A second engine noise can be heard mixing itself in with the regular beat of a fishing boat making its way along the horizon. The noise comes from a car driving down the track, stopping briefly at the car park. Almost immediately it sets off again through the soft sand, heading straight towards the sea. It is clear to those watching that the open car without doors is a four-wheel-drive vehicle. It almost looks like one of those American Jeeps. Engine racing, the odd truck ploughs its way through the loose ground, turns around and repeats the exercise. Again and again the car sticks to the same route leaving circles and furrows in the wet sand. After a while it turns around and disappears over the rocks, up the track and drives back on to the main road. Decades later, people will say that you can still see the tracks of the first ever Land Rover on the beach of Red Wharf Bay, but only if you look hard enough.

War and Peace

A brief interlude on the first all-wheel-drive vehicles

The history of four-wheel-drive is as old as that of the car itself. Right from the start, people attempted to get the engine's power to all four wheels. There were two reasons for this: First of all, it appeared to promise greater top speeds. On the other hand, it also promised-and this was of considerable interest to the military – better traction and made vehicles thus equipped, able to operate independently of the road network. It is not surprising therefore, that most of the early development work was undertaken on behalf of the world's various armed forces. One must also differentiate between all-wheel drive- and off-road vehicles. This chapter will highlight only the major developments in all-wheel-drive technology and leave the history of all-wheel-drive transporters to others.

The first patent for a mechanical four-wheel-drive system dates from 1898. Its inventor was a Robert E. Twyford from Pittsburgh, Pennsylvania. However, it was not until 1905 that he actually converted theory into reality. The honour of being the first car to have four-wheel-drive fell therefore to the Spyker 50HP from Holland which was first launched at the Paris Show in December 1903. Designed as a racing car, it was constructed by Joseph Laviolette and had an 8.7 litre straight six engine. Its power was transferred to the wheels via a three-speed gearbox and a two stage reduction 'box (giving a total of six forward and two reverse gears) and a centrally mounted differential with propshafts to both axles. Thus equipped, the outer wheels were prone to turn more quickly through curves than the inner wheels. Although this technology gave the 50HP considerable tractive advantage, the then current materials technology was not up to the stresses and test runs usually ended prematurely as a result of material failure and mechanical breakdown. So this remarkable vehicle remained a one-off, although a few 50HPs were later built and delivered with four cylinder engines.

When driving a car was still the sport of gentlemen. The Spyker 50HP was equipped with four-wheel-drive and a straight six engine. Today, the vehicle is in the Rosmalen Automobilmuseum in Holland

Meanwhile, four-wheel-drive technology was making further progress. In 1908 Otto Zachow from Clintonville, Wisconsin patented a steerable front-wheel-drive system he had developed with his brother-in-law William Besserdich and built into an off-road vehicle which he had built himself. A further development of this vehicle appeared in 1911 with the firm's logo FWD (Four-Wheel-Drive). As early as 1912, the firm presented its first four-wheel-drive truck finding immediate favour with the American military. This forward control truck is worthy of mention because a vast number of them was built between 1914 and 1928, sources referring to more than 40,000 units. Consequently, it can be referred to as the first volume production all-wheel-drive vehicle. A number of vehicles made their way to Europe and some were built under licence by English manufacturers.

Concurrently, the interest in more manoeuvrable, lighter off road vehicles was also increasing. In 1923, the Americans built a Model T Ford with four-wheel-drive and huge balloon tyres. Apparently its off-road capabilities were satisfactory. However, performance and capacity were well below what people had been hoping for.

Tempo G1200 equipped with two engines. Its traction system was later employed in the Citroen 2CV Sahara in 1958

One of the most challenging tasks facing four-wheel vehicle manufacturers is preventing the drive assembly from grounding out on firm ground. As the inner part of a curve is shorter than the outer part, the outer wheels travel further in a curve than the inner wheels. Consequently, it is imperative to be able to prevent transmission wind-up and to keep wear down to a minimum (see page 21). The classic and most simple solution to this problem is part-time front-wheel drive with which the first volume production four-wheel-drive car was equipped. This was the Black Medal Scout Car and was built by the Japanese manufacturer Kurogane first appearing in 1935. Equipped with a front-mounted V-2 motorcycle engine, independent suspension and an open body, it was capable of carrying five people. Most of the production run of 4,800 vehicles up until 1940 was taken by the Japanese military.

Unconventional four-wheel-drive technology was being used by the Hamburg-based German firm Tempo to find an extremely effective solution to the problem. The G1200 was equipped with two 600 cubic centimetre 19bhp two stroke engines with flanged gearboxes that each drove wheels which were attached to the loadbearing central tubular frame and gave the option of front-, rear- or all-wheel-drive. Additionally, an all-wheel-steer system, which could be disengaged from the rear axle, made the G1200 amazingly manoeuvrable and gave it a turning circle of only seven metres. Thanks as well to its patented and triple independent suspension front and rear, the Tempo was able to take all the honours at many off-road events. Justifiably, the vehicle is feted as the most capable of the off-road vehicles of the thirties.

Between 1936 and 1944, 1,335 examples of the marque were built and exported to places as diverse as Australia, Brazil, Chile and Iraq. The Swedish Army alone received 985 vehicles up until 1944. The Germany army never procured the G1200 because it had a two-stroke engine. Not so the Horch 901 off-road car (1935-1942), the 108 (1937-1942), the Mercedes-Benz G5 from 1937 onwards, and the Stoewer Models R180 and 200 Special from 1936 which did see active service. Power came from a range of four, six and eight cylinder motors, the vehicles were all approximately four metres in length and came equipped with all-wheel steering, a five-speed gearbox and three differential locks. The Volkswagen, which was introduced in 1938, also appeared in a number of all-wheel-drive variants. As well as the four-wheel-drive Schwimmwagen (types 128 and 166), 1941 saw the introduction of a command car with

Long version of the Horch 108 1a from 1940. As with the Tempo, the Horch was equipped with spare wheels which were carried amidships on ball-bearing-equipped mountings which prevented the vehicle from grounding

Below: Mercedes G5 Alpine Watch vehicle in operation

Crawling Beetle: The Type 87 Command Car. In Germany there are only two that are known to have survived

bar grip tyres on a raised suspension using the famous Beetle body (Type 87).

On board the Beetle, the engine was the well known 1.1 litre Boxer unit which put out 25bhp. The drive was transferred to both axles, each equipped with a differential lock (whereby the front-wheel drive was operated via an off-road gear). Including other VW all-wheel-drive derivatives, approximately 600 command cars were built until 1945. A further two Type 87s were built for British occupation forces after the war.

The list of early off-road vehicles would not be complete without including a milestone of all-wheel-drive technology which celebrated its fiftieth birthday in 1997. This was the Universalmotorgeraet (Universal motorised apparatus) better known by its shorthand "Unimog". The vehicle had actually been developed during the war by the aircraft manufacturer Albert Friedrich as a contract for the Erhard and Sons Gold and Silver factory in Schwaebisch Gmuend. The prototype was completed in March 1946 and thirteen months later this was presented to technology experts. Finally, it was presented to the public in Frankfurt in August 1948. The Unimog was based on principles which were both simple and functional. The first production version was driven by a 25bhp, four-cylinder Mercedes diesel engine. Directly above the engine was a narrow two-seater cabin with a canvas roof. Thanks to its short wheelbase, high ground clearance, portal axles, a six-speed gearbox with two reverse speeds, plus a two-stage reduction gearbox, part-time four-wheel-drive, and differential locks on both axles, its off-road capabilities eclipsed anything that had been built before it. Numerous power-take-off points, a wide range of bodies as well as its one tonne carrying capacity, made it the master of all tasks. For this reason and also because of tax, it was classified as a tractor. From Autumn 1948, the Unimog was built by the Boehringer Brothers in Goeppingen. Lack of adequate manufacturing capacity led to the Mercedes-Benz factory in Gaggenau taking over production from 1951 onwards. Since then, over 300,000 vehicles have been built.

Tractor, pickup, off-road vehicle or truck? The Unimog can do everything and is still built today

It can be seen therefore that elements of the evolution of the off-road vehicle came from all over the world. Four-wheel-drive was not something that can be ascribed to the Americans alone which is often made out to be the case. The Americans however, were the most consistent in its development and the common effort of the American automobile industry culminated in the Jeep which was launched in 1941.

The Jeep came about as a result of an invitation to tender put out by the American General Staff in Camp Hollabird, Maryland. The need was for a multi-purpose vehicle with four-wheel-drive. On July 5 1940, the corresponding construction specifications for

a four-wheel-drive quarter-ton truck were sent out to 135 potential manufacturers. Although many of the recipients were interested in the highly lucrative contract, only one, the American Bantam car company from Butler, Pennsylvania made the September 23 deadline for the first prototype. The military top brass were so impressed by the Bantam Reconnaissance Command HP40 penned by Bantam Chief Designer Karl Probst, that two other companies, Willys Overland from Toledo Ohio and Ford in Detroit, kept to the original Bantam design with their own prototypes. As Bantam were not capable of manufacturing the required numbers quickly and cheaply enough, the General Staff insisted that Willys and Ford concentrate on one vehicle in order to simplify production. The choice fell on the Willys. Ford's model was little different from the vehicle that had been produced in Ohio.

The open steel body of the vehicle which was now known as the Jeep (GP stood for General Purpose) rested on a robust ladder frame chassis and leaf sprung beam axles. Power delivery came via a three-speed gearbox, transfer- and additional reduction gearbox. Front-wheel drive was a part-time arrangement. Under the bonnet was a side valve 2.2 litre in-line four cylinder petrol engine putting out 61bhp which came from an earlier Jeep model, the Whippet, and was capable of propelling the Jeep along at just under 60mph.

Thanks to reliable technology and its service with allied forces, the Jeep became a household name the world over and was the basis of many future developments. The Land Rover was no less successful. There is however a significant difference. Although the Land Rovers would later be conscripted into military service, it was designed and constructed for solely civilian purposes in a period when the Jeep had just passed its peak production period. Between 1941 and 1945 exactly 647,927 Jeeps were built, a figure that would never be achieved again in such a short space of time. Land Rover's triumphal march occurred during peacetime and over 1.7 million have been built thus far.

The model for all light off-road vehicles: The Jeep. On the left, the original Bantam Reconnaissance HP40 of which 2675 were built. Next to it and above the version built by Willys and Ford

How Shortages became a Virtue

The strength of a brand and the origins of the Land Rover

Maurice Fernand Cary Wilks (1904-1963), the father of the Land Rover

The history of the world's most successful off-road vehicle began in the spring of 1946 with one of the most unpleasant storms accompanied by lightning and hail to hit the Midlands and which caused a great deal of damage. Maurice Wilks' own farm at Blackdown Manor in South Birmingham had not been spared either and some huge trees had been snapped in two like matchsticks.

Wilks was a graduate engineer and technical director of the traditionally-minded car manufacturer Rover. On the day after the storm, he started clearing up the damage around his garden using a World War II Universal Carrier with an open rear load bay which he had acquired from military surplus. Using this vehicle and the help of a forestry worker he had enlisted specially for the task, Wilks was able to clear up the tree trunks from his land in a relatively short space of time. His helper the forestry worker was so impressed by the Universal Carrier and its large eight cylinder engine that he proposed a deal to Wilks. The tracked vehicle for an old chainsaw and a used Willys Jeep. Wilks had, apart from this one tree-felling opportunity, hardly had a chance to play with his "spoils of war" and did not hesitate for a minute to say yes. Little was he aware of the significance of this piece of horsetrading.

The sign of the Viking Longship. This was the logo of the Rover Car Company during the Thirties

In order to understand the history of the Land Rover, with which Maurice Wilks was implicitly involved, it is important to understand something of the history of Rover. Rover came about in 1884 almost as a by-product of the cycle manufacturer Starley-Sutton which itself had been founded in 1877 in Coventry.

In 1902, the company produced its first motorised vehicle, a motorcycle. As with most manufacturers of two-wheeled vehicles, the company did not wait long before concentrating on the much more promising car business. As early as 1904, the Model 8HP appeared with its air-cooled one cylinder motor and what was a first in the motor industry, a central tubular frame. Rover continued to produce bicycles and motorcycles until 1924. The late twenties saw the company in dire straits thanks to its weak financial position: Stagnating sales, poor management leading to over-production in the anticipation of future sales and not least, worrying quality problems left the account books looking decidely questionable and people started asking questions about the future of Rover itself.

1929 saw a real turnaround in the fortunes of the company in the person of one Spencer Bernau Wilks. Maurice's thirteen year older brother was not unknown in the business. A trained lawyer, he had joined Humber Hillman after World War I and with his good ideas and management style had become Vice-Chairman. Thanks to these excellent references, he was very quickly appointed one of Rover's managing directors. He also brought along his brother Maurice who had also worked at Hillman where he had shown considerable technical ability. Although it would be a few years before the company was in the black again, it was thanks to these two brothers that Rover managed to put the crisis behind it.

One of Maurice Wilks' first tasks was the development of a simple, small car for which management had just given the go-ahead as it appeared very promising. The Scarab (which was so-named because of its size) was a four seater with a folding roof, powered by a 7bhp rear-mounted air-cooled two cylinder engine. This configuration was all the rage at the time and yet it had no luggage space. Nevertheless, Maurice also gave the tiny car individually suspended rear wheels and Spencer estimated that the vehicle would cost an unbelievable £89. However, the board changed its mind and the project was cancelled.

Spencer Wilks' plans to both tighten up and modernise the model range appeared to hold greater promise. In 1932, two new vehicles, the Family 10 and 10 Special were introduced as part of a new model range. The latter vehicle also came with a freewheeling gear, separating the motor from the gearbox, which made gear changes much easier. The new 10s came in the nick of time as the mistakes made during the late twenties led to the first financial losses in 1931 which culminated in a severe loss of £25,000. Without Spencer Wilks at the helm, Rover would have had difficulty surviving the following months. With his own peculiar long term view of things, he was able to turn the struggling company around completely bringing it back into the black. For this, he was promoted in 1933 to Managing Director. In August, Rover introduced further new models: a modern four cylinder with a three-bearing crankshaft and overhead valves proved to be very popular with the car-buying public. Spencer Wilks' staged rationalisation programme and his maxim "Quality first" started to bear fruit. In the following years, Rover was able to establish itself as a marque for those with tastes above the average, using the slogan "One of Great Britain's finest cars" and even became known as "The poor man's Rolls Royce". By 1939 and the outbreak of World War II, profits had increased threefold over 1933 and Rover had achieved the number one spot in the number of new registrations.

The little Rover Scarab never got beyond the prototype stage. The engine was in the rear, the grille at the front was false

There were however, new priorities facing Britain's industry. The crisis in Europe had worsened as a result of Hitler's expansion plans. For this reason, the British military authorities had been working on a rearmament programme from 1936 onwards. Supplementary and camouflaged production facilities, the so-called "shadow" factories had been prepared to produce strategic components. Britain's largest motor manufacturers including Austin, Daimler, Rootes, Vauxhall and Standard were asked to get involved in the rearmament programme. Consequently, Rover took up production in a factory in Acocks Green financed with government money in the summer of 1937. The next step came in 1939 when the government decided to construct the Bristol Hercules radial engine using Rover factories. To this end, a further factory was built to the north of Solihull and came on line in January 1940. Additionally, repair and maintenance work on aero engines finally put paid to any vehicle production. May 1940 saw the last pre-war Rover car leave the production line. A few months later the German air force attacked Coventry and laid the city and its factories to ruin. Solihull however, managed to survive the bombing raids unscathed.

During this period, Rover was stepping into new technological territory. This involved the design and construction of jet engines. Frank Whittle had been working on a similar aircraft propulsion system for a number of years. The decisive final step getting the engine fit for mass production had been beyond the capabilities of Whittle and his firm, Power Jets. Nevertheless, there was a race on to get the engine into action as it was feared that the Germans were also working on the development of a jet engine. Indeed, Junkers and BMW were actually at a similar stage. However World War II had virtually ended before the technically very advanced, if somewhat unreliable Me 262 jet came into service. British engineers working on the Gloster Meteor jet were working under similar pressures to get their own twin-engined fighter into the air. Ultimately, both aircraft entered service far too late to have any major impact on the course of the war. In the early war years, it was obviously impossible to guess how things would turn out and haste was very necessary.

One of the Rover engineers involved in the project was Tom Barton and he would later play a major role in the development of the Land Rover. The then 26-year old joined Rover in 1941. He had only just received his call-up papers from the British Army and they gave the railway engineer the choice of either going to an army staff college or working on the jet engine project. Barton chose not to join the army and joined five other engineers at Rover. One of his colleagues was the slightly older Gordon Bashford who would also figure in the Land Rover story later on. After test bed runs with the Whittle engine had demonstrated that the principles of reverse flow by which the engine had been designed, were causing it to run inefficiently, the design was fundamentally changed. By the end of 1942, and codenamed B26, the first jet engine, with a straight, aerodynamic flow which more than equalled the performance of Whittle's original design, was up and running and the ministry gave the Rover version the go-ahead for mass production.

While this decision naturally spurred on the team at Rover, it led to a major disagreement with Whittle who felt he had been pushed aside and taken advantage of. Nevertheless, the B26 was built, albeit not by Rover where capacity was again lacking. In 1943, pragmatism took over and the complete project, including all design work and the completed test bench mules, were handed over to Rolls Royce which had expressed an interest in the project and also had enough capacity available to take it on to production. One might suggest that without the Rover turbine, Rolls Royce would never have got involved in this technological field. Today, and given the current parentage of Rover Group it is a matter of some irony that Rolls Royce Aerospace is in a joint venture with BMW. The British company currently accounts for some 25% of the total market for both military and commercial jet aircraft engines.

Meanwhile back in 1943 Rover, which only ten years previously had looked bankruptcy in the face, was now a successful engine manufacturer and in return for the B26 project was given the task on behalf of Rolls Royce to develop the Merlin, the powerful twelve cylinder petrol engine with 27 litre capacity and 700bhp which powered the Spitfire into a diesel variant for tanks called the Meteor. Engineers were given the option to go with the turbine to Rolls Royce or to stay with Rover. Barton and Bashford who, by now had become good friends, decided to stay at Solihull and work on the diesel monster. The swap with Rolls Royce did not mean the end of gas turbines at Rover. After

Tom Barton at his drawing board. Before joining Rover he worked for a railway company

the war, the company would get involved with them again, albeit in a completely different way (See pages 41-42).

In the spring of 1945, and with Germany defeated, Rover was, as in the thirties, confronted with a potentially fatal situation. Military production had come to an end and the buildings of the once upmarket manufacturer were either empty or in ruins. Only the Solihull plant remained largely intact and for this reason Rover moved its headquarters there in the summer of 1945. In theory, one should have been able to start up where production left off however, Rover, like every other manufacturer, had major problems to deal with. The lack of production capacity and shortages of both financial resources and steel, made it virtually impossible to put new ideas into production. Consequently, it was decided to stick with the construction of the pre-war P214 limousine. The country as a whole was in a precarious position as far as the economy was concerned. Rationing was still in force. Most people had greater things on their mind than whether or not to buy a medium size limousine, especially as petrol rationing, which prevented people from travelling any distance by private car, was still in force. During this difficult period, people were particularly sceptical about the future. As far as the Wilks brothers were concerned, one thing was clear: If Rover wanted to survive, then it had to produce a vehicle which matched the requirements of the time. Although work was in progress on a follow-on version of the eight year old P2 – the new model was to have the independent suspension which had been developed by Gordon Bashford – it would be three years until the new vehicle called the P3 appeared on the market in 1948. The completely new P4 would also take four years to develop. There was clearly a need to fill the gap and keep the company's head above water.

In early 1945, and with the end of the war in sight, Maurice Wilks and his technical department had already dreamed up a compact car which could be offered at a favourable price. The ideas crystallised into the construction of three prototypes with open and closed bodies. In keeping with the simplicity of the project, the vehicle was called M1 (the M standing for minimal). It was kept as simple as the Scarab had been fifteen years previously.

Unlike the Scarab, the engine in the M1, a 700cc 28bhp four cylinder affair was mounted in the front between independently suspended wheels, while the rear beam axle was equipped with modern coil springs. Another novel feature was the pressed aluminium body frame. The progressively-styled bodywork with integrated headlights

The M1 was a smart, compact car with two doors and four seats. Only economic circumstances prevented it from going into production

was also manufactured from aluminium with which Rover intended to get around the shortage of steel. Indeed Spencer Wilks is thought to have planned a complete model programme around this British version of the Topolino and to produce 5,000 models a year. That things did not quite work out this way was down to two facts: The first was that the country needed hard currency. The priority of the period was made only too clear by the government's battle cry, "Export or die". Companies producing export goods were dealt with more favourably by the government and could count on receiving more steel.The board felt that the M1 was too small to be considered for the export market. Consequently the M1 project was put on the shelf. The second reason had four wheels and made its debut in early 1946. It was a small Ferguson tractor and rapidly became a best-seller causing Rover to rethink its plans. There was a very real need for agricultural equipment and one of the Ferguson's strengths was its ability to run a range of implements from power take offs. The "Little Grey Fergie" as it was nicknamed because of its grey paintwork, was manufactured by the Coventry-based Standard Motor Company. Standard had similar difficulties to Rover in the immediate post war years and was only running at full capacity thanks to the Ferguson. Consequently, people in Solihull began thinking about their own vehicle that would be multi-functional, particularly as it appeared that the plan to build a primitive farm workhorse for the agricultural industry could be turned into a full production vehicle quite quickly. Although this signalled the end of the M1 project, there was still no consensus among the Rover technicians as to what would be an ideal utility vehicle concept. They did not want to make a tractor. Rover was, after all, a car manufacturer with a good reputation which they wanted to keep. As well as that, it seemed to make no sense to put up a direct competitor to the Ferguson which had already proved its worth. There had to be another way.

A year later, Maurice Wilks had the idea that was to save the company while on the isle of Anglesey. This was where his family regularly spent their holidays staying in a hotel on Red Wharf Bay on the north east coast of the island. As well as this, the route to Anglesey offered excellent conditions for Maurice to test new prototypes: The diverse Rover "test route" with its curves, gradients and slopes through the Welsh mountains could almost have been designed with testing in mind. Wilks liked the island so much that he bought a large piece of land with a cottage called Tros Yr Afon on the edge of the village of Dwyran in the south east corner of the island.The Wilks family used a range of Rover vehicles as family cars and also the Jeep he had acquired in the deal with his Universal Carrier. Sometime in the summer of 1946, the Jeep went with the family on holiday to Wales. The war veteran coped with the journey without as much as a hiccup and also performed well on the island. It was no problem taking a small sailing boat down to the beach. Wilks was impressed with the vehicle's handling and reliability. However, on the coast the relentless march of time moved rather more quickly: Salt water and the raw environment very soon transformed the green-painted Jeep to a different hue.

By now it was spring 1947 and Spencer Wilks was visiting his brother on the island. He noticed the rather delapidated state of the Jeep and asked his brother "what are you going to do when it packs up?" He answered succinctly, "I'll just go out and buy another," at which point Maurice realised that there

The farmers' darling "Little Grey Fergie" immediately proved a great success. The tractor firm Massey Ferguson, based in Coventry is now one of the biggest in the sector

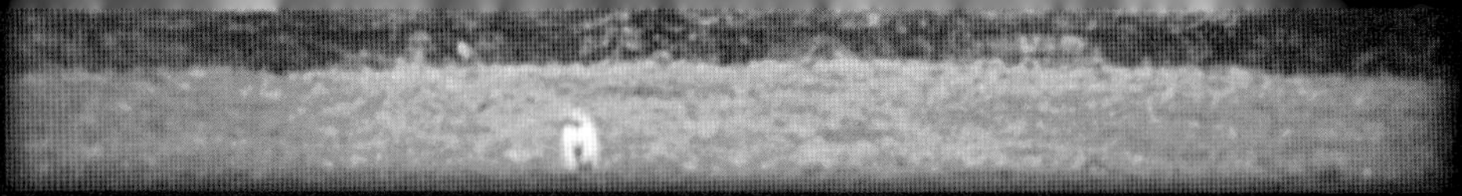

From the Wilks' family photo album: After severe storms, Maurice Wilks takes the Universal Carrier out into the garden...

...and drags the toppled trees off his land. Finally he swapped the tracked vehicle for a chain saw and a Jeep

A few weeks later Wilks took the Jeep (note the matching number plate FWD), boat and oars to Wales. In the background is a Rover P2 Saloon

Holidays on Anglesey: For the next trip the boat is loaded onto the Jeep at Tros Yr Afon called affectionately by the family "Trosy". The married couple from next door, the Owens, Mrs. Wilks (with hat) and the children all muck in...

...and Mr. Owen secures the load

In Red Wharf Bay the dinghy is launched into the water. Less than a year later, Wilks will return to this spot with another off-road vehicle

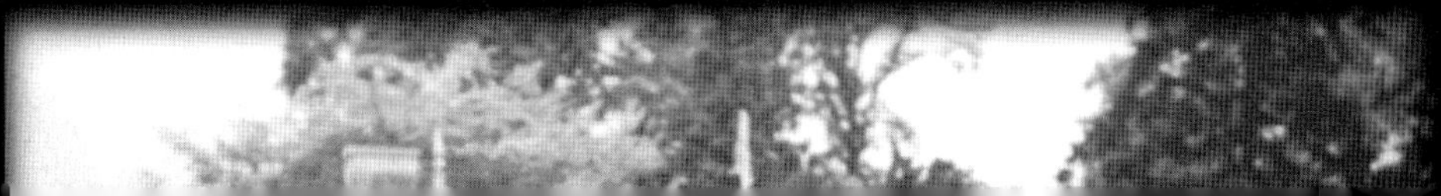

Finding Tros Yr Afon takes time. Towards the end of the seventies, the Anglesey cottage was sold off

was nothing similar available. He had discovered the gap in the market and found what Rover was urgently looking for. That very week, he ordered Rover technicians including Tom Barton and Gordon Bashford to come over to Wales and demonstrated to them what the Jeep could do. Barton remembered his exact words "We are going to build a car like this! A Rover for a farmer so that he can get everywhere and do everything. It will be a universal Land Rover."*

Back in Solihull, the team got to work immediately. Wilks handed over the development to five departmental heads demanding, "Make it simple!" Thus the project was carried by only a few people, each of whom had responsibility over very specific areas. The man in charge of engine development was Joseph Drinkwater, Barton was responsible for the drive assembly and Bashford was responsible for the vehicle's backbone, the ladder frame. Initially, he purchased two scrapped Jeeps from a breaker's yard, which were then taken apart. The chassis were used as a temporary starting point for the prototype. This was a relatively simple procedure which goes some way to explaining why the wheelbase of the original Series I Land Rover is identical to the Jeep.

This does not mean however that the Land Rover is merely a derivative of the Jeep. Rover had had no experience of building a four-wheel-drive vehicle, nor had they a suitable chassis for such a vehicle. Consequently in the initial stages, they merely relied on something that was tried and tested. Time and money obviously had something to do with it as well. The modified ladder frame chassis now took what was in the Rover parts bin: The 1.4 litre four cylinder petrol engine and the freewheel

** The project designation Land-Rover was used right from the start and fitted so well that it stayed. Since 1980 the name has been written without the dash; for the purposes of continuity, the modern spelling will be used throughout.*

The two prototypes were built using Jeep chassis like this

gear from the old Special 10, gearbox, differentials and drum brakes from the 1937 P2 limousine as well as numerous parts from other Rover models. Pretty much the only thing that had to be developed from the ground up was the planned reduction gearbox which would give the vehicle low and high speed capability. In the off-road setting it would give the vehicle more power at lower speeds and also increased its engine breaking capability on slopes. The reduction gearbox positioned to the right of the gearbox required relocating the propshafts which also meant that the leaf sprung beam axles had to be completely redeveloped. Because of their off-road stability and articulation capability, they are particularly suitable for off-road vehicles and are still used today unlike in most road-going cars which are largely equipped with independent suspension systems.

The Land Rover was originally conceived with permanent four-wheel-drive. The resulting permanent transmission wind-up between the two axles – caused by the lack of a central differential – made the vehicle difficult to handle on the road. This dramatically increased tyre wear and put increased strain on the whole drive train assembly. Finally, the steering characteristics in 2WD and 4WD were so different that it could lead to problems for the inexperienced driver. Off-road, this is of little concern because low speeds and the odd spin on loose ground take up any transmission wind-up that occurs. To deal with these problems, the first Land Rover was equipped with the freewheel by which the front wheels could be disconnected from the drive assembly. All the driver had to do was pull a lever in the dash board.

All that remained to be done now was to develop a suitable body. When it came to which material was to be used, the consensus was for aluminium. The Land Rover had been conceived as an export model and the steel requirement would have been increased accordingly. There was an important argument however that spoke in favour of using aluminium. It could be

A few weeks after the project's launch, the first Land Rover was complete, albeit unpainted in the workshop

A place in the middle. The centre steer Land Rover could not be beaten for its outright simplicity. Nevertheless, it was equipped with side indicators and footsteps to make getting in and out easier. The windscreen at this stage was still a one-piece affair

A prototype being tested in the field. Certain similarities to a tractor were par for the course

shaped without the expensive pressing equipment that was required for steel bodies. In the same vein, it was decided to opt for a squared-off body because this would be easy to manufacture. Once the matter had been discussed, the engineers simply got on with the task of building it. The actual manufacture of the first bodies took little more than a couple of days as they didn't come about on the drawing board. Instead, and in a manner which is hard to believe these days, approximate measurements were made and the panels for the bonnet, wings and side walls were cut accordingly. This would not have been so easy with steel. Despite these obvious advantages, "Duralumin" had one disadvantage. It cost three times as much as steel which itself wasn't exactly cheap and came to be the most expensive element in the vehicle's development.

It turned out to be a fortuitous decision to stick with this material because the bodywork which was both light and barely susceptible to corrosion contributed in no small way to the Land Rover's good reputation. The fact that the vehicle was developed in a time of shortages and was a totally functional machine with little or no creature comforts were strengths which contributed to its position as a milestone in automotive history.

By the summer of 1947 and less than two months after the project had been given the go-ahead, the protoypes were standing completed in the factory. One was a chassis without a body, the second was painted with spare green and grey paint. Their most noticeable feature was the centrally-mounted driver's seat, which was designed to make it easier to produce without having to do separate versions for export. The Wilks brothers informed the Rover board about how much progress had been made and the board gave the go-ahead for limited production on September 4 1947. Construction began immediately on a series of 48 pre-production vehicles that differed in a number of areas from the two prototypes. It was now no longer possible to use the Jeep platform and a new chassis was constructed by Gordon Bashford. As steel wasn't available in the lengths required, the chassis rails of the ladder frame were initially constructed out of two shorter lengths which were welded together. Beneath the slightly different bodywork, the more torquey 1.6 litre four cylinder petrol engine from the P3 60 (which appeared in February 1948) replaced the 1.4 litre engine. As well as this, the isolated, central driving seat was considered to be too tractor-like and also took a lot of getting used to. Consequently it was replaced by a more conventional left- or right-hand drive layout. Additionally, modern telescoping shock absorbers were mounted. Other than these items, the technical specification was pretty much identical to the prototypes. Finally, the rounded-off front wings were dropped in favour of more squared-off versions as these would be easier to manufacture. Concurrently, the Land Rover team created their own off-road test track in a small wood on a neighbouring piece of land adjacent to the factory, (with emphasis on the off-road). Somewhat nostalgically, Tom Barton thinks back to those extremely busy days, "Every spare minute went into the Land Rover project and it was loads of fun!"

Using an old Cromwell tank, winches, axes and shovels, the team dug out some deep ditches, created steep inclines, mud

The 1948 Rover P3 differed only in certain optical details from its predecessor and provided the Land Rover with a number of its innards

lakes and rocky test tracks. Everything was done to provide the right testing environment for the all-wheel-drive Rover. The pre-production vehicles would be tested to the limit using this test facility. The longevity of the Land Rover did take away from the smoothness of the ride. According to Tom Barton, "After the first few test runs we decided to leave the vehicle as uncomfortable as it was for a very simple reason, you wouldn't go mad off-road." The prototypes were scrapped at the end of the test programme. Along with some Rover technicians, Maurice Wilks took a pre-production vehicle out to Anglesey for a rehearsal run. The Jeep's capabilities on the island were well known to everyone and the Land Rover was put through an extensive series of tests on the farm's terrain and across the island's sandy beaches. A few days later the expedition returned highly satisfied, reported to the Rover board and the Land Rover was signed off for production. The Rover board did not attribute a great deal of importance to the Land Rover. At the end of the day, it was little more than a stop gap measure and it was intended to cease production as soon as the P4 appeared.

Up until this point, no risks had been taken. Investment had been kept to a manageable minimum and after the final run out, it would have been a small matter to convert the production line back to manufacturing saloon cars. Just exactly what kind of horse the owners had in the stables would only be realised later on. Despite the crack on pace – barely twelve months had passed by since Maurice Wilk's brilliant idea – it still wasn't possible to get the exhibition models ready for the Geneva motor show. The next opportunity after that would be the Amsterdam Motor Show in April 1948 and by then everything would be ready. The Land Rover, Europe's first civilian off-road vehicle, was ready for its first public appearance.

What's done is done: Rover employees Bill Whale and Timber Woods with the prototype on the newly built test circuit. Solihull's Jungle test track still exists today, though potential customers will be glad to hear that their own Land Rover would not have been forced through it

From Farm Nag to Workhorse

An off-road vehicle conquers the world: Series I

In the history of the British automobile industry, 1948 is a memorable year. Three vehicles, a sportscar, a compact car and a four-wheel-drive caused an international storm and renewed the good reputation that cars "made in Great Britain" had enjoyed before the war. In the first instance we are talking about the Jaguar XK120 whose elegant bodywork had been designed in only two weeks (!) by William Lyons. It nevertheless served as "adequate packaging" for Jaguar's real new arrival, an in-line six cylinder with twin overhead camshafts which had been developed for the future limousine, the Mk VII presented to the public in 1950. Another potential car of the year was a Morris produced by the Nuffield Organisation and created by Alec Issigonis (the legendary designer of the Mini). Seeing the light of day in October, this was the Minor. The Minor stayed in production until 1971 by which time over 1.3 million of this English Beetle had gone down the lines. The glory however belonged to another no less exciting motor vehicle on April 30 1948. When the doors to the Amsterdam Motor Show opened, some members of the Rover board were visibly nervous as they waited on the stand not knowing how the public would react to the vehicle. They had never built an agricultural vehicle before. There was however no doubt as to the strength of the design. It was the customer who was to have the last word.Would he recognise and value the usefulness of this functional off-road vehicle, which fitted no marketing segment and certainly did not match the traditional image of Rover? Maurice and Spencer Wilks did not have to suffer too long. Pleasantly surprised and at the same time feeling that their belief in the vehicle had been borne out, they noted the interest of the crowds. The Land Rover was one of the sensations of Amsterdam, the trade press immediately swamped the vehicle, as did the first interested future customers as they pressed forward to get a look at it. Even while the trade fair was still running, the company took hundreds of orders and transformed the odd nervous smile into broad satisfied grins.

The quickest production car of its time: The Jaguar XK120

Popular plain cooking: Morris Minor

However, not even the Wilks' nor Tom Barton, not to mention any of those involved with the project would have believed in their wildest dreams that the Land Rover would still be in production fifty years later in 1998 – an almost biblical age which has been reached by very few other vehicles.

The model presented to the public at Amsterdam had a wheelbase of 80 inches, an open pick-up body as well as a galvanised chassis to protect it from rust and a front bumper which was welded to the chassis. The exhaust terminated on the left-hand side beneath the passenger door. The utilitarian cabin was equipped with rather spartan seats with tiny padded backs which were attached to the rear bulkhead. Despite the extremely narrow profit margin that was to be earned

This picture of the Land Rover at the 1948 Amsterdam Motor Show which had been lost for fifty years was finally located by intrepid Dutch Land Rover enthusiast Richard de Roos in March 1998 – just in time for the anniversary

One of the first Pre-series production Land Rovers with a galvanised chassis. Today "Huey" can be found in the Gaydon Heritage Museum after being discovered by Rover staff still being used by a farmer in Warwickshire.
It was bought back and restored. The bumper and seats depicted in these photographs are not those which were originally fitted

An early Rover publicity film – in the fields with suit and tie

Less than a year old, the Land Rover is given the Royal seal of approval. Here King George VI tests the Land Rover

Sceptical faces:
On the right, Rover boss Spencer Wilks and Rover manager Geoffrey Savage pictured next to an export model

Even Sir Winston Churchill, then still Prime Minister supported the British motor industry and accepted the adoration of the crowd in a Land Rover. This scene dates from July 15 1948

Using Power Take-Offs on the nose, rear and in the middle of the Land Rover, it was possible to operate a wide range of apparatus

from the £450 sales price, the men in Solihull considered offering such items as the doors, plexiglass sliding windows, canvas tonneau, passenger's seat, spare wheel mount and even a starting handle for the engine as optional extras. This idea was abandoned fairly early on however, as Rover didn't to want to offer its new baby with too little in the way of equipment. Additionally, it was hoped that by making the package attractive enough, more buyers for the Land Rover could be found. The favourable price was achieved by being able to avoid the relatively high sales tax of 33 % that was slapped on vehicles in Great Britain at this time. This was done by virtue of the vehicle being classified as an agricultural tool. This claim was underlined by the modest way in which the vehicle was presented to the general public. The power take-offs remained optional extras however. A hand throttle which was hidden under a flap and mounted underneath the middle seat regulated the output of the engine in order to operate the PTOs. Early on, the production line was on an adjacent track in the P2 production building and initially only twelve completed vehicles left the line each week.

During the first official year of production only 80 vehicles were completed. 48 of these were the pre-production model and 32 the first production variant. Land Rover sources tell a different story however, indicating that none of the 1947 pre-production vehicles were classified as volume production. The first calendar year of volume production was therefore according to Land Rover, 1948 and in that year the total volume production amounted to 1,758 units. The British Motor

1 four cylinder engine
2 radiator
3 air filter
4 12v battery
5 front axle differential
6 steering track rods
7 clutch/brake
8 accelerator pedal
9 handbrake
10 reduction gearbox lever
11 spring release for front wheel drive
12 gearstick
13 gearbox
14 reduction gearbox
15 drum brakes
16 rear prop shaft
17 rear axle differential
18 drum brake
19 tank
20 direct PTO hand-throttle
21 belt drive
22 PTO propshaft
23 chassis
24 body
25 tailgate
26 PTO
27 belt drive
28 towing bracket
29 silencer
30 exhaust tail pipe

Industry Heritage Trust records a total of 3,048 Land Rovers for 1948. The production vehicles differed from the pre-production vehicles in a number of details. The exhaust now came out at the right rear corner and the bumpers were modified. In order to ease repair, they were simply bolt-on affairs. The freewheel gear lever was also moved to a position in the driver-side footwell next to the transmission tunnel and was operated by a ringpull. The galvanised chassis was also dropped and instead, 'waxoyl' was sprayed

Special versions were offered right from the start such as this tractor unit equipped with oversize tyres parked in front of the Solihull factory

Land Rover production. On the left, a rolling chassis complete with engine and recently sprayed bulkhead (in the background is a Rover P3) On the right is the production track

Export success: A load of left-hand drive vehicles about to leave the factory

into the insides of the chassis box sections through specially drilled holes. Finally, the body panels were now manufactured from Birmabright aluminium as the first field tests had demonstrated that Duralumin was vulnerable to corrosion if it came into contact with certain kinds of fertilisers.

Even before the Amsterdam show Maurice Wilks had stood down his team of engineers from the project. They had done their work and would be needed for new projects. All except one: Tom Barton was picked out to keep his eye on the Land Rover. After all, he was one of the engineers who had been involved in the project from the start and knew the Land Rover inside out. With the benefit of hindsight, Barton named other reasons, "I was a railway man and not an automotive engineer like the others. That made me pretty dispensable as far as car development was concerned." Initially, Tom Barton reckoned that after the planned two year production time he would have to find another job. However, as we now know, things turned out rather differently.
A few months after the launch, Solihull

On the edge: The Land Rover could cope with extreme angles of 45 degrees

received some unwelcome mail. Some Land Rover customers had discovered fractures in the chassis of their vehicles. The cause was soon established. The ungalvanised chassis had proved to be less sturdy than the original version so Barton ordered that more steel panels were used in the construction of the chassis to stiffen it up. Alongside these improvements, he also experimented with new forms of welding. Tests had established that if the sheets used to manufacture the chassis rails were given toothed edges, the resulting weld was much stronger. This technique was immediately integrated into the production process.

At the same time production of the P2 limousine also came to an end, and its successor was ready to go. Even before the P3 was officially anounced to the general public, the Land Rover was already using the modified transmission from the P3. The spare production capacity was now being taken up in order to produce Land Rovers as demand increased. Around the same time, Rover also offered two specially-equipped vehicles with a compressor and welding

equipment. They were however, very expensive and there was little demand for them. Nevertheless, the supreme, all-round capability of the Land Rover encouraged numerous smaller companies to offer tools and equipment which could be operated with it.

At the time it was built, the Series I Station Wagon was not a best seller. Today, it is much sought after.
Most examples of the type were exported

Along with the special vehicle options, Rover launched another body variant in October 1948 as a reaction to the unexpectedly high demand for the vehicle. This was the seven-seater station wagon. The closed body was based around a wooden-framed, aluminium-skinned body which Rover contracted out. A one-piece front windscreen, winding front windows and lockable doors were some of the features of this model. Nevertheless, the Station Wagon was not a success. As it was obviously a passenger-carrying vehicle and not a commercial, the sales tax soured the success by forcing the vehicle price up to £959, a figure which was almost double the cost of its smaller brother which had already risen to £540. When production stopped at the end of 1951, only 641 station wagons had been built.

The sales figures of the open version continued to rise unabated and the failure of the Station Wagon was soon forgotten. By the end of 1948, 1,758 vehicles had rolled off the production line. By the summer of 1949, Rover was able to report the production of a further 6,000 Landys (the name by which the vehicle was lovingly referred to by their owners). By October 1948, Rover was already exporting to over 68 countries and in early 1949 the company was forced to take on more staff in order to increase capacity.

Almost a car:
The hard-top Landy differed by having exterior door handles, winding windows and seven seats. Each part of the two-piece rear door window could be raised and lowered separately. The spare wheel cover had to be ordered separately, too

Wherever one travelled, the Land Rover was there first. This model was parked in Port Moresby, New Guinea

Argentinian Land Rover fans: Head of State Juan Domingo Peron and his wife Evita in front of the presidential palace Casa Rosata in February 1949

Versatile vehicle indeed: Down under the Land Rover quickly became popular. Here the motoring correspondent of Country Life magazine (with the pipe) and the Australian head of sales on the passenger seat check over the first vehicle to be imported into Australia by Sydney-based Greenville Motors

In 1949 Rover introduced the "Auntie" P4. The version shown is the "100" built in 1960

In the meantime, the P4 appeared to be the long-awaited beacon of hope for the company. Technically, it was closely linked to the P3 and in contrast to its rather barge-like appearance, rested on a boxframe chassis. Even so, the P4 came equipped with a fully synchromesh gearbox and "Auntie" as the P4 became known because of its rather austere and frumpy appearance, represented a seamless link to the pre-war traditions that Rover stood for: Namely solid, conservative vehicles with above average build quality. People appreciated this in the UK and the P4 rapidly gained in popularity. One could have assumed that by now the Land Rover had done its duty and served the company as a stop-gap measure. However by now no one was thinking of stopping production any more as Land Rovers were selling like hot cakes. Over 1949/50, production increased from 12,395 to 16,795 units.

The Land Rover's overwhelming success can in part be attributed to the market situation at the time. Initially, there was no competition whatsoever, although Austin had tried to grab a slice of the cake for themselves in 1949. In 1948, the British Army's Fighting Vehicles Research and Development Establishment (FVRDE) announced the requirement for an off-road vehicle coded FV 1800 and offered Britain's biggest motor manufacturers the chance to compete for the contract. Rover declined. Tom Barton recalls, "When we politely turned it down, Austin jumped at the plans."

The Austin Champ which appeared in 1949, differed in some important areas from the Land Rover: Its wheels were individually suspended on an x-shaped chassis and it was powered by a 2.7 litre four cylinder motor with 75bhp. The engine, designated B40, was manufactured by Rolls Royce and was part of a family of engines which were standard in British military vehicles. Of particular interest was the four-speed gearbox with an additional crawler gear. The vehicle (which was not fitted with a conventional gearbox) was equipped with a device which enabled the car to move forwards and backwards in any gear. It was equipped with a part time four-wheel-drive system but no reduction gearbox. With all its complicated technology however (and the fact that the body was steel), the Champ weighed a stately 1600kgs. Although Rover had turned down the FV 1800 project, there was still interest in a potentially lucrative military contract and the attempt was made to mount the the standard engine in the

The Competition: An Austin Champ battles with mud. The Champ was not bad off-road, but the Land Rover won through in the end

Land Rover. However, as the B40 engine was too long, the chassis had to be lengthened by one inch between the front axle and the bulkhead. As Solihull had neither the space nor the time to work on such an extensive conversion, Rover contracted the work out to Hudson Motors Ltd in London to convert a number of Land Rovers and fit the B40 engine. The limited production run of 34 vehicles can be recognised not only by their longer front bodywork but also by the raised bonnet. Using an 81inch and a Series I Land Rover, the company entered into trials at the beginning of 1950. Although the Austin and the 81inch Land Rover conversion were more powerful than the standard Land Rover, this advantage was cancelled out by an unfavourable weight penalty in the Austin and the 81inch Land Rover over the front axle. In short and much to Tom Barton's joy, the standard Series I came out of the military mud trials a clear winner. This guaranteed the Land Rover further success as the army immediately ordered Land Rovers by the hundred. And the Land Rover was bestowed with military honours. In the same year the Land Rover was put into UN service in Korea and six years later finally became accepted as the standard British military vehicle. The new military orders and ever-increasing demands for the civilian versions – in 1951 two thirds of all Rover products were now Land Rovers – forced Rover to expand production capacity again at the beginning of the fifties. During this period, the Land Rover was also continuously developed. The original version which had only been offered with a canvas roof and frame now appeared rather too basic and accordingly from February 1950, a steel hardtop was added to the options list. Four months later this was also offered in aluminium. Initially, the hardtops came

Land Rover goes diving. Obviously the engine and electrics had to be sealed against water ingestion. Air came via a snorkel

without windows in order to maintain the tax benefit. Additionally, there were some small cosmetic changes. From May 1950 onwards, the headlights were moved from their position behind the radiator grille to an open position actually within the grille itself. New standard indicators on the front wings replaced the trafficators mounted on the sides of the windscreen which had been an optional extra. Two years later, changes were also made to the rear body sides and wheel arch panels which were now taken down lower.

As part of a rationalisation programme, the Landy received the gearbox from the reworked P4 and from October 1950 the freewheel gear was replaced by a dog clutch as the freewheel arrangement had not proved totally reliable. From now on, the Land Rover had the option of four wheel drive in high range; in low range, all four wheels were automatically driven. A new engine which had been developed for the P4 also gave the Landy a more modern powerplant. In August 1952, the two litre engine putting out 52bhp replaced the 1.6 litre motor from the P3 60. Although the new engine had only two extra horsepower, it produced more torque at lower revs than the previous incumbent. Finally, the following model year saw a major change in the door handles. The canvas flaps covering the door handles which were mounted on the inside of the doors, were finally removed as mounting them on the outside had made the flaps superfluous.

Who would have thought it possible: In 1952 the 50,000th Land Rover came off the production line. However they were just the first...

The original version with open body, aluminium panels, spare wheel in the rear load tray, a deep petrol tank underneath the passenger seat (if that had been ordered), plexiglass sliding side windows and fully extending doors

While work progressed on the four-wheel-drive vehicles and the limousines, Rover continued to work on the development of the gas turbine engine. The decision had been made in November 1945 by the board to try this engine in vehicles. Under the technical leadership of engine designer Frank Bell and a young man called Spen King, the development work started the same year. King was a nephew of the Wilks brothers and had only recently joined Rover from Rolls Royce. Together with Bell, King constructed a test bed and in May 1948, one month after the launch of the Land Rover, they presented the first ever car gas turbine engine. Further versions followed and in the winter of 1949 the Rover board gave the go-ahead for its installation in a car. Gordon Bashford was given the task of re-engineering a standard P4. This is how the open two seater Jet 1 with a centrally mounted gas turbine that put out 230bhp at 40 000rpm(!) to the rear wheels via a reduction gearbox and a cardanshaft came about.

The first tests of the gas turbine appeared quite promising and in 1952 Rover founded a separate department called Rover Turbines Ltd which somewhat underlined the importance that Rover attached to the

Two 86inch Series Is with the newer exterior-mounted lockable door handles and later headlights belonging to the Dutch KLM airline

The fruits of success: After the record run in Belgium, the Jet 1 is pictured on the factory site with Rover staff

project and which also separated the enterprise from the vehicle manufacturing division. During this period, Bell left the project and King continued on his own. Towards the end of June 1952, he took the Jet 1 over to Belgium and got the vehicle up to 153mph which was the first world record for a gas turbine powered car. Suitably motivated, King developed other gas turbine powered cars, the T2, T3, T4 that caused real comment in the engineering world.

The extremely high fuel consumption as well as the excessive development costs were not kept in check. This was a situation that Graham Robson aptly described in his book "The Rover Story" as hardly being shameful as "no-one else has managed to do it either." The last performance of the gas turbine engine was in the Rover BRM racing car which took part in the 1965 Le Mans. That race sealed the end of Rover's involvement with gas turbine engines. Spen King turned his attention to a new project, the development of a luxury Land Rover, the Range Rover (see pages 86-99).

While the four-wheel-drive and the passenger cars were continuously developed, Rover also experimented with a new form of tractor in 1952. Under Barton's direction, two prototypes were put together

The future of the automobile? Actually not as Rover's gas turbine, (pictured is the 1956 T3 with a fibre glass body and four wheel drive) proved to be the wrong approach. Nevertheless, parts of the suspension would later find their way into future Rover car models

with a chassis that was very similiar to the Land Rover's, albeit one of smaller proportions with four-wheel-drive and four-wheel steering. As well as this, it was possible to adjust both the front and rear axles in order to give the tractor an additional 25 centimetres in ground clearance in less than 30 minutes. Additionally, the rear wheel steering mechanism could be locked. The diesel powered vehicle certainly looked spectacular. It was clear to many that the spectacular success of the Ferguson tractor had certainly stirred something in Solihull and volume production was considered. Nevertheless, some remained sceptical as development appeared too complex and consequently too expensive. During testing on the jungle track, one tractor rolled over backwards after the driver forgot to lock up the rear-axle steering and this sounded the death knell for the project.

Still, the Land Rover continued to be developed. At the start of 1954 it grew. The model 86, as the name says, had a six inch longer wheelbase. One of the major criticisms of the Land Rover right from the start had been its relatively small load area. Increasing the rear body by three inches not only demonstrated Land Rover's response to customer demands for a bigger load bay, but also caused the marketing men to breathe a sigh of relief. Underneath the bonnet, changes were also afoot. Prior to the introduction of the 86, the four cylinder engine had been replaced by another completely new engine from the P4 60, albeit with the same capacity and performance. This engine, somewhat simplified in comparison to the saloon, had a different carburettor, a grey iron cylinder head and lower compression. Additionally, the 86 received stronger brakes. The exposed, internal door handles were replaced by new ones which were mounted in the doorskin itself, a feature which has since become synonymous with Land Rover.

Parallel to the production of the 86inch, Rover launched a lengthened pickup with an enclosed cabin called the 107. Apart from the

Among the more unusual projects at Rover during the fifties were these two tractors with four-wheel steering and articulating axles. They never made it beyond the prototype stage however

unsuccessful Station Wagon, it was the first decisive new model in the six year history of the vehicle. During those years the Land Rover had been the recipient of a number of improvements. The lengthened leaf springs served not only to increase the wheels' suspension travel, but also in conjunction with the lengthened wheelbase, improved the roadholding as well. Apart from that, the technology remained the same. A few months previously, the drum brakes had been enlarged. The traditionally draughty doors and windows were given rubber seals. Interior ventilation was improved via two fresh air vents directly beneath the windscreen. (A single vent flap had been an optional extra since 1949.)

As well as these, newly designed and more easily readable instrumentation, an open parcel shelf and sprung seats were introduced to give a modicum of comfort which could be added to via an optional comfort pack including roof lining, door liners, and carpet. However it wasn't only the 107's body and interior changes which made the Landy more pleasing to behold – for the first time, the Land Rover could be had in blue or grey as well as the standard green.

As early as Autumn 1954, Rover launched a seven seater 86inch. The closed 107inch also came with a station wagon body with four doors, and elongated, so-called Alpine lights in the roof edging sections. The vehicle could even be ordered with ten seats. Furthermore, the various models could now be ordered with a tropical roof. Specially designed for hot weather climates and tropical regions, it maintained a through

World Rover: Confidence in the product marked the Rover approach to marketing the Land Rover. Above: the new 86 with new instrumentation and rear lights. From 1955, the side windows were made of glass

Right: Off-road, people started organising competitions around the Land Rover

flow of fresh air directly above the cabin roof. The standard, light paintwork also reduced the chances of the cabin heating up too quickly. As both versions were classed as passenger carrying vehicles, they also were also eligible for British sales tax. However thanks to the use of cheap aluminium, the increased price was kept within reasonable limits.

The 86 and 107 were built for two years before themselves being replaced by even longer versions, the 88 and the 109. The reason for the revision was quite simple. Rover planned to introduce a diesel engine. Independent engine specialists Turners of Wolverhampton had offered a loud, clanking diesel engine in 1953 with two litre capacity from two cylinders and pumping out 40bhp. In order to demonstrate the reliability of its diesel, the company sent a three litre version to Nairobi as part of its Operation Enterprise.

The Turner engine nevertheless remained a rarity. However, somebody at Solihull must have liked the diesel. It may have had something to do with lower fuel consumption, longer service intervals and reasonable maintenance costs. Initially Rover did not want to build the engine itself, but after fruitless discussions with Standard-Triumph and Ferguson, Rover decided to develop its own engine anyway.

The preconditions for the new engine had been created by Barton by bringing the 88 and the 109 to market and a few months later in June 1957 the powerful diesel, a grey iron block four cylinder with overhead valves,

The new 107 Pick-up heralded the launch of a new model in 1954. Barely a few months passed before the station wagon was being christened. Two Australian importers do exactly that

PRENSA
VENEZUELA
AUTO-AGRO C.A. CARACAS
VENEZUELA

a capacity of 2.1 litres and 51bhp arrived on the market. Only the "short" 107 station wagon remained unchanged and without the diesel until the introduction of the Series II which appeared in 1958. In the first ten years of production over 210,000 Land Rovers had been built and of these over 70% had been exported to over 150 countries. Meanwhile, the off-road vehicle had travelled to the most far-flung corners of the world. Hardly a country remained where it had not left its tyre tracks. Arguably, it was during these heady days that the Land Rover legend was really created. It was now up to Solihull to move the vehicle on.

Equipped with a tow ball or just as a means to go shopping, even on the tarmac, the Land Rover was a multi-purpose vehicle

In the mid fifties, this 107inch Land Rover from Caracas in Venezuela crossed the Southern American continent. Here, it is in Cotacachi, Ecuador

The Station Wagon satisfied customer demands for more storage capacity and seats. The 107 also impressed people with its off-road ability. These pictures were taken by then Land Rover importer in Germany, Hamburg-based Tempo (see pages 79-81)

Thanks to the robust nature of its chassis, the Land Rover was the perfect base for special bodies. This is an ambulance equipped to carry four stretchers and left is a mobile greasing station

A one off: This coach with a glass fibre body on the 107 inch chassis was built by the coachbuilder Mulliner as a private contract for an eccentric couple

The Swinging Sixties

All about chart breakers and dropouts: Series II, IIA and their derivates

Series II models awaiting transportation from the factory car park at Solihull

Comparison: Rover engineers also took the opposition to task. In this case, an early Toyota Land Cruiser

With a series of improvements across the board, the Land Rover went into its eleventh year of production. The Amsterdam show was once again chosen as the stage for the Land Rover which was now known as a Series II. The launch was to be in the spring of 1958, exactly ten years after the launch of the first model which was now known as the Series I. Up until this juncture, the differentiation between the two marks had not been necessary. It had been a while since Jeep and Land Rover were the sole players in the four-wheel drive market looking for custom. Ever since its launch in 1952, the Toyota Land Cruiser (one wonders where they got the name?) had enjoyed increasing sales success. In the mid-fifties, Jeep extended its own range on the American market, while at the same time expanding into overseas

They would never be this clean again. Landys being prepared on the edge of the track between P4 limousines

Landy drivers had waited a long time for this. The filler cover was now on the right

manufacture in Japan, Brazil and Argentina, and in England, Austin tried to re-enter the market with the Gypsy which was however less than successful. This later model, which was technically very different from the previous vehicle which had been built until 1956, was taken off the market in 1968. The Land Rover had proved too strong. While the original Series I had been a veritable spartan in a time of austerity, now the Land Rover styling department really set to work. Designer David Bache was however particularly careful with the product. As far as he was concerned, thanks to the good basic design, other than mere adjustments to some minor details, very little change was actually necessary. Consequently the bodywork barely changed, growing in width by 1.5 inches. The body now had a barrelled-out side extending through to the front wings which were given a more squared-off appearance. New doors, front indicators and new rear lights were also part of the visual changes. Additionally, the Series II had painted window frames. For the first time the Land Rover had an exterior petrol tank filler pipe; now it was placed immediately behind the passenger door and not under the right front seat. In order to ease filling from jerrycans, a metal funnel could be pulled out from the tank filler pipe. Further concessions to driver comfort were made with regard to the interior of the Series II: Larger pedals and new seats were more comfortable. In the 109, the driver's seat was even adjustable. The choice of colours and accessories was also extended. Free-wheeling hubs for the front wheels which could be locked and unlocked, were made available. This spared the drive shafts and the differential which now didn't turn unnecessarily when the front wheel drive was disconnected on-road. Obviously, it was advisable to lock up the freewheeling hubs before venturing off-road.

Piggyback to Dover: Eight 88s for export to the European continent

Still going strong: This is what a well-cared for two and a quarter litre petrol engine looks like

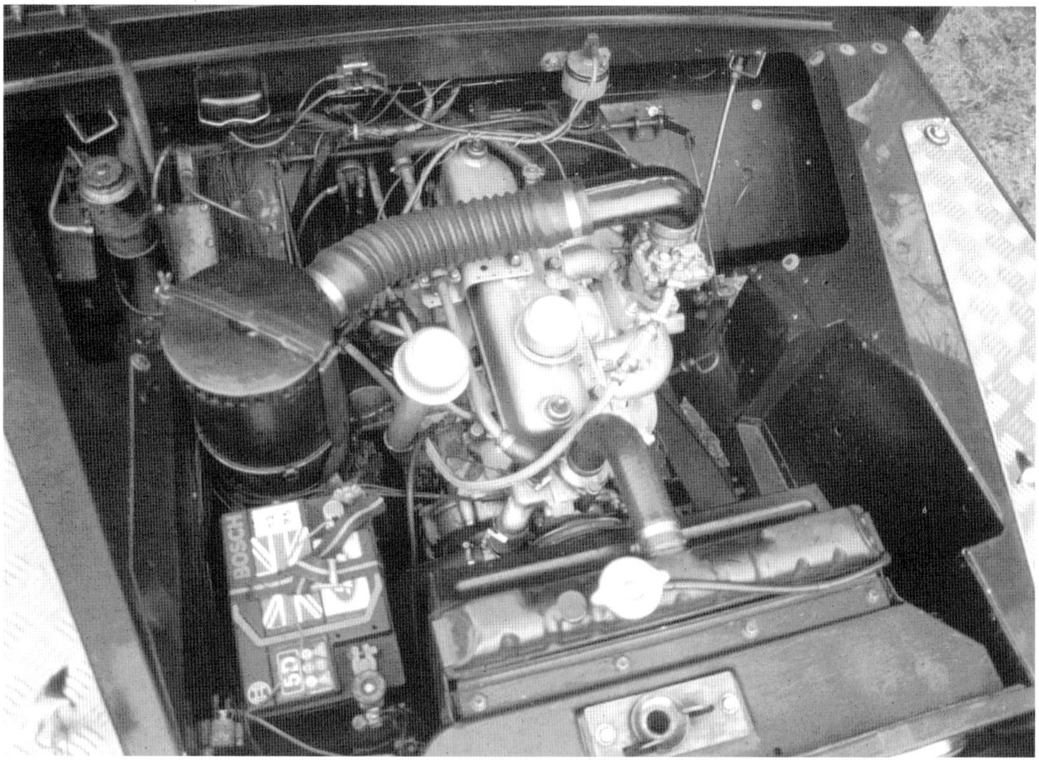

A vision of straight lines and corners: The cockpit of the Series II. Windscreen wiper motors and ventilation pipes were still open to the world. The doors were held by catch rods

One of the technical changes on the long wheelbase version was a reduced turning circle of 1.5 metres. Thanks to the greater width, there was also more space in the wheel arches and the track increased by 1.5 inches. Nevertheless a 13.7 metre turning circle was still rather large. Perhaps the most important technical development was an increase in cylinder capacity of the petrol-engined variants to 2.3 litres (or more accurately 2286cc) Linked to this was a hike in performance of 40%, from 55bhp to 77bhp (at 4250rpm); the maximum torque figure increased by 23% from 137 to 168Nm. The Land Rover still was not exactly a fast motor and 70mph was still the maximum attainable speed. The Rover remained a sales success story. In 1959, the 250,000th rolled off the production line and each year up until 1961 saw record leaps in production. In 1958, the price of a Series II had reached £640. The long wheel base now cost £730. The diesel engine option cost a further £90.

From this position of considerable strength, Rover entered into discussions with other British manufacturers towards the end of the fifties to further consolidate itself. Unfortunately, the negotiations led nowhere. Rover was however, looking well beyond British shores. Even Jeep manufacturer Willys

These photos prove it happened: For comparison purposes, the Land Rover engine was implanted into a Jeep

was considered as a possible partner, whereby actual partnership was not considered rather an agreement to share technical knowledge. Bringing the world's two best off-road vehicles together to create an even better product was a logical step. People have almost forgotten that in 1958 a Jeep was fitted with a Land Rover engine and gearbox. Rover engineers were not exactly enthralled by the project. Even today, Tom Barton refuses to acknowledge that it actually happened. A few years later mergers would occur that would almost bring the British motor industry to its knees (see pages 70-71). Rover was still basking in the success of its off-road vehicle which was now beginning to attract more and more military orders. In 1959, the Australian Army chose the Land Rover as its standard utility vehicle and in 1960, Switzerland followed suit.

In 1961, the capacity of the diesel motor was brought into line with that of its identical petrol sibling. Consequently the Series II now become known as the IIA. Apart from that, the introduction of the newest model was pretty much a cosmetic act with only minor optical detail changes. Eleven additional horses and an increase in torque of 22Nm, recognised the demand by diesel owners for ever more power. Now putting out 62bhp and 140Nm, the diesel was still slower than the petrol-engined variant.

Note the Land Rover details inside the Jeep: gearstick, instruments, steering wheel, pedals, and factory plate

The Landy also made a name for itself among the camping and caravanning fraternity. Numerous suppliers and converters were involved in special camping conversions

In North Africa the Land Rover was already well known. For expeditions requiring a lot of equipment, the Land Rover is the perfect tool

On the right: As the numbers of Land Rovers increased so did the size of the fan club. In front is a model with a de-luxe bonnet which initially was only available as an option. In the driving seat is a youthful Roger Crathorne later to become head of the Land Rover Driving Experience

In September 1963, the fifty-nine year old Maurice Wilks suddenly and unexpectedly died. He was buried in the vicinity of his beloved farm on Anglesey. The inscription on the gravestone mentions the Rover Gas Turbine, but not the Land Rover which was then barely fifteen years old. Little did people realise that the Land Rover would become a classic.

Maurice's brother Spencer left the company in 1961 after thirty years service. He died in 1971 aged 80. His successor was William Martin-Hurst. 1962 saw the introduction of the 109 twelve-seater variant, once again for tax purposes: Vehicles that were used for commercial purposes were still not eligible for puchase tax. Clearly the rear bench seat made the Land Rover a passenger car. Finally someone within Rover management discovered the ruling that vehicles equipped with more than eleven seats were classified as buses and therefore tax exempt. Thus the ten-seater became a twelve-seater. Now instead of costing £1,293, the 109 cost £950. Instead of two benches each seating two and facing forward, three

557 JUE

Land Rovers were often trusted with unreasonable loads, but they were equal to the task

individual seats were fitted. It is fair to say that it wasn't that comfortable in the back. Three years earlier in 1959, Morris launched the Mini. In terms of the genius of the concept, the Mini was very similar to the Land Rover. Today, both are classics. Towards the mid-sixties, the Beatles were conquering the music world and defining a generation. Meanwhile, on board the Land Rover, very little had changed. Yearly sales reached new heights however in 1965/66 with record figures of 48,000. It would be wrong to assume that the Land Rover team merely rested on their laurels. Almost simultaneously, two projects came to fruition which would represent new stablemates for the Land Rover. One was the Forward Control (FC) and the other a specially developed version for the military which had had its weight pared down; the half tonne Land Rover or as it was later to become known, the Lightweight. Appearances might have had something to do with the fact that the Forward Control was not exactly a success. Although it was the most practical of vehicles it was also incredibly ugly. As the designation implied, the engineers had taken a 109inch chassis, mounted the engine between the chassis rails and placed the cab directly above it. This saved space which resulted in extra capacity in the load bay. Although the first forward control prototype first ran in 1960, it was September 1962, before this mule-like load carrier had its premiere at a commercial vehicle show in London.

Tom Barton claims the idea for the Forward Control as his own. The way he tells it, he took a few Land Rovers to a military presentation in 1957/58 at which other manufacturers were also demonstrating their products. One of these was an off-road truck, which appealed to the top brass. Armed with a "we can do that", Barton went back to Solihull and set about the task. It should be mentioned however that Jeep had already been working on a similar concept in 1956 and had even given it the same name: The Jeep Forward Control. Was this a coincidence? Whoever had the idea first, it was true that the standard Land Rover's weight limit was limited even in its long

Another one off: This mini Landy from 1960 was a Rover special production for the son of the King of Jordan. Before the vehicle was delivered, Peter Wilks, a nephew of Maurice and Andrew allowed his four year old son Andrew to take it out on a test drive

Sometimes you can go a bit too far: Here it is clear to all and sundry that this Land Rover is on the limit

wheel-base Pick-up variant. During the fifties, work had already been done on a truck variant with a 129inch chassis and a wider track. It had even got to prototype stage with five being built. It was arranged conventionally with the engine, cab and load bay arranged in that order. The new forward control had double the carrying capacity of a pickup with the same wheel base: 1.5 tonnes. As well as this, the forward control was technically quite advanced for its day. Drive train and gearbox all came from the standard Land Rover.

To cope with the increased capacity, the axles were strengthened. Engineers were particularly proud of the fact that a high percentage of the FC's components had come straight from the parts bin. Only one in four parts could not be provided from componentry already being manufactured. In 1962, a Forward Control cost only £1,015. For your money you had a truck with an open load tray measuring 3.14 by 1.61 metres. The FC's overall length of 4.9 metres represented an 18 inch increase over the standard 109inch. Nevertheless, even when unladen, the FC weighed 1.9 tons forcing the rather weak two and a quarter litre petrol engine to work extremely hard. 1966 saw a revised model brought onto the market. Externally, the only visual difference was that the headlights were now mounted lower down. Underneath, a six cylinder engine with 2.6 litre capacity and 83bhp from the Rover P5 range took up service. From now on, the four cylinder engine was only available in the FC for export markets. Additionally, the diesel engine was also made available, however this

The forward control concept freed up more load-carrying space. Pictured is the first prototype

The FC was practical but a real ugly duckling

Thanks to the forward position of the driver's cab, it was possible to dramatically increase the size of the load tray. The FC could also be had with a bus body. The FC's cab was virtually identical to the standard Land Rover's

was an engine that was truly inappropriate for the FC. Advertising brochures proclaimed the new six cylinder as the standard engine and in order to fit it, the wheelbase was extended by one inch and the track grew by four inches. It was however, all to no avail. The FC flopped. In 1973 production was finally stopped. Later versions built around the same concept would only be offered to the military (see pages 93-94).

Nevertheless, it is particularly this kind of Land Rover that the dyed in the wool fans revere, perhaps because they are so rare. In addition, their popularity might have something to do with their greater ground clearance (ten inches) and the optional 9.00x16 tyres that could be fitted (standard equipment was 7.50x16) made them better performers off-road. There were some minor differences between the Series IIA and the IIB. The latter had a 110inch wheelbase instead of the former's 109 and also was equipped with the 2.6 cylinder side-valve petrol engine for the home market. In the IIA, only the export versions received the more powerful six cylinder while the domestic market had to make do with the standard four cylinder 2286 petrol engine. Not surprisingly, finding the power to make the IIA move forward was always a struggle. The last major delivery went out to South Africa in 1973. By that time, a fairly meagre 5,500 of

The FC's predecessor: Only five examples of this conventionally laid-out truck were built towards the end of the fifties

this unloved utility had been built by Solihull. Rather higher production figures were achieved with the second Land Rover cousin, the Lightweight. This special version came about as a result of a British military requirement in the sixties for an off-road vehicle which weighed less than 1133 kilos or 2500 pounds – the military designation 'half-tonne' referred directly to the vehicle's payload capacity. The term 'Lightweight' has proved to be more durable with enthusiasts both in and outside of the military. The limitation placed on the vehicle referred to the underslung weight limit of the army's standard transport helicopter, the Wessex.

There were also numerous special versions of the FC such as this rare crewcab version. Fire brigades across the world were particularly keen to have the FC like this version built by Carmichael and Sons from Worcester

Amphibious Land Rover: This one off special came about as a result of an Australian Army tender and was equipped with a six cylinder engine and twin fuel tanks

Parts of the bodywork and even the tyres of this amphibian were filled with buoyant foam and enabled it to float. The small side hatches could be folded over to facilitate entry and exit

On the slipway: In the water the vehicle was propelled by a propeller which was positioned just in front of the rear axle. Its top speed was a paltry four knots

was another vehicle that was born in the sixties arising out of a demand by the Special Air Service for a heavily-armed light attack vehicle. The name came about because of the paint used to camouflage it for its desert role-pink. Apparently pink was the most difficult colour to differentiate from the desert background. Automotively speaking the vehicle was hardly different from a civilian 109 other than uprated suspension. The FFR (Fitted for Radio) Land Rovers were equipped with a 24 volt electrical system shielded against electrical interference. Also favoured by the military were twin fuel tanks each holding 11 gallons, which were positioned Series I fashion, underneath the driver and passenger seats. They also had to be filled separately. Via a lever, the driver could switch from one tank to the other while underway

Hard Nut: This Shorland armoured car with a machine-gun turret and bullet-proof tyres can be seen at the British Motor Industry Heritage Trust Museum at Gaydon. The Shorland name was an amalgam of the two Belfast-based companies, Short brothers and Harland who combined to design the armoured body

and the fuel display was automatically switched over from tank to tank. In order to improve the battle-readiness of these fighting Rovers, many were equipped with larger than standard radiators and twin drive belts. From the front, a military Land Rover can invariably be recognised by the heavy duty double front bumper, with corner pieces known as bumperettes.

Meanwhile, back on civvy street, a new 2.6 litre, six cylinder engine which had come from the Rover P4 and the Forward Control made the headlines in April 1967. It was retained for service in the long wheel base models. An improvement of six horsepower and a tiny increase in torque from 168 to 173Nm does not seem very impressive when viewed in today's terms. However in 1967 it was all about prestige and the fact that the

The tracked Centaur combined the tracks of a Scorpion Light Tank with the front end of a V8-equipped Land Rover

Designed for maximum mobility in Scottish bogs, the ungainly looking Cuthbertson conversion could be steered via the front tracks

For the man for whom four-wheel-drive just wasn't enough. Specialist manufacturer Cunningham built this 6x6 from an 88inch base vehicle

Who called the Land Rover a tractor? Specialist converter Roadless Traction Limited produced this outlandish vehicle for forestry work. This vehicle is pictured being put through its paces on Solihull's jungle track

Land Rovers had to prove themselves underground as well, like here in an African copper mine

On the right track: As well as this road-rail version with an extended track, there were also trolley Land Rovers with large toothed wheels

Production Values: April 1966 saw the 500,001st Land Rover roll off the production line. Surrounded by Rover directors, Rover boss William Martin-Hurst is sat behind the wheel, next to him is Tom Barton. On the far right, the man with the glasses is Gordon Bashford

maximum torque figure was achieved at lower revolutions (1500 instead of 2500) meant that the engine was much smoother. Additionally, the engine only cost £60 more than the standard four cylinder. The introduction of the six cylinder also led to a complete revision of the interior which was now rather more luxurious. The position of the dashboard instrumentation was also restructured and now included a water temperature dial. The push start button also disappeared, for the first time a mere turn of the key would bring the Land Rover to life. 1967 also saw the introduction of negative earth electrical systems to the Land Rover. At the time, Land Rovers, along with most other vehicles, were positively earthed – a feature which has caused many an innocent restorer of a pre-1967 vehicle to tear their hair out in frustration. 1968 also saw a significant revision to the 109 Pick-up. Strengthened axles and larger wheels from the 110inch Forward Control brought the capacity of the vehicle up to a ton.

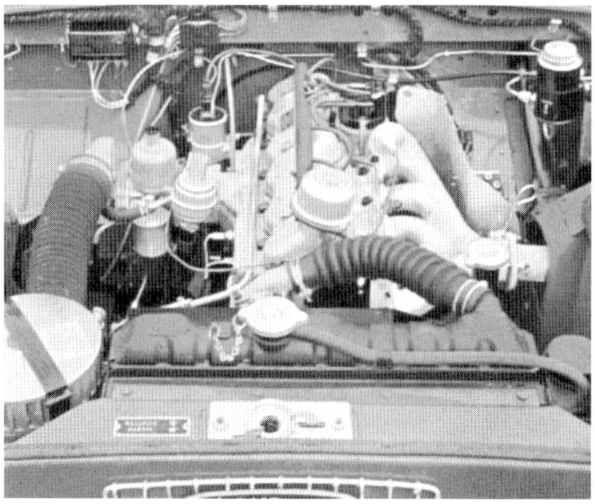

Sixpack: The 2.6 litre side valve engine was introduced in 1968. Unfortunately it offered little more power than the four cylinder engine

Simultaneously the Land Rover was given a facelift. As a result of various legal requirements (particularly those coming out of the United States) the headlights were removed from their traditional home in the radiator grille to the wings. To those versed in Land Rover lore, this development has been one of the most difficult to explain. As many pointed (and still point) out, the headlights belong in the middle. As far as enthusiasts of the marque are concerned, Land Rovers that were built after 1968 are definitely No. 2 in the pecking order. The original reason for mounting the headlights centrally is as simple as it is obvious: No matter what damage the wings suffered, the headlights could always be relied on to function because of their protected position. Tom Barton was also thinking along similar lines when the design of the split windscreen was worked on. If, during transportation, a screen was damaged, replacing one half screen instead of a single piece screen would always be cheaper.

There are some very rare examples of Series IIA models which merely had their headlights screwed into the front wings and project out from the wing somewhat. Every once in a while one might find a Land Rover with four headlights. One that made it on to the cinema screen was the IIA used in the film "Four Weddings and a Funeral".

The Series IIA was built until October 1971. Apart from the revised headlights, everything on the IIA, and in particular the four cylinder models was kept pure and simple just as the aficionados liked it. Bare metal in the interior, small instruments, no glove compartment, the starter button somewhere in the middle, uncomfortable seats, the simplest of door locks and exposed screws and bolts everywhere. Naturally the engine could be fired up using the time-honoured method of a starting handle. This reminder of the early days of motoring is particularly good at collaring the enthusiasm of those who are unfamiliar with the marque. Climbing aboard a Series I or II is like stepping into an automotive time tunnel. 1971 represented another record year for Land Rover with sales of 56,600 vehicles. Production amounted to 56,674 vehicles. If one looks at the sales of the classic Land Rover and its current equivalent, the Defender, this figure has only been beaten by including those vehicles produced by Santana in Spain. The Discovery however, has managed this feat twice (see page 141).

1968 saw the headlights move into the wings. On the left is one of the first examples with the new face

Roll-out: After having built 549,311 vehicles, the Series II/IIA was pensioned off in 1971. It was exported to over 200 countries

Stormy Weather

How Land Rover survived the crisis years: Series III

Jaguar Rover Triumph Ltd
This version of the world famous Land-Rover range is 88 in. wheelbase, 7-seater station wagon Neg no. 15129/3

United in strength? After the mergers of British manufacturers in the sixties, it ended in misery. This is a press release from the late seventies

No man is an island. The same can be said for Land Rover. There is no doubt that Tom Barton's engineers had enjoyed a great deal of success over twenty years and the sales figures certainly spoke for themselves. Unfortunately, Land Rover did not speak for itself . It was after all "only" a part of Rover and if you believe some of what the old timers had to say, Maurice Wilks always set great store by the fact the Rover car company built cars first and the Land Rover was almost ignored. However, Land Rover never made a loss – the same could not be said about Rover cars, particularly during the seventies. True, Rover was a company with great strengths, but it was still unable to escape the whirlpool that overtook almost the whole of the British automotive industry during this period.

Before World War II, the British Motor Industry was as prosperous as it had ever been. There were a good dozen well-known car manufacturers. In 1952, there was a first wave of rationalisation which saw Austin, Morris, Wolseley, Riley and Vanden Plas merged together in the belief that unity brought strength. Thus unified as the British Motor Corporation, the amalgam took Jaguar on board in 1966. Rover had itself taken over Alvis in 1965 and would have been able to survive a while longer, but as money for investment into future car models and for the necessary plant expansion simply was not there, the company was forced to look for a strong partner. In 1967 the merger came about with the bus and commercial vehicle manufacturer Leyland which had itself just taken the sports car manufacturer Triumph under its wing. For Rover, this constituted the end of 90 years of commercial existence as an independent. If you include Rover's forerunner Starley-Sutton, then Rover is the oldest British car company. The merger of the two largest car manufacturers in 1968 created British Leyland. Apart from a few isolated marques, all major manufacturers were now united under one roof. However, instead of the anticipated greater commercial weight, the conglomerate suffered from the differing philosophies of the individual companies it had amalgamated. As well as this, constant strikes crippled production. On occasions, the production lines stood still for weeks on end. The money that was needed to modernise the plants was also not there. In the time it took to say British Leyland, the traditionally fine reputation that British cars had enjoyed, was destroyed. Rover suffered from the association with this period for a long time afterwards. In 1975, the state had to intervene and save this massive concern from bankruptcy. Meanwhile the erratic course continued. Three years later, the conglomerate was divided up creating the unlikely sounding manufacturer Jaguar-Rover-Triumph alongside Austin Morris and for the first time, Land Rover became a separate operating company within the group.

Apart from the financial crisis, Land Rover had long since ceased to be the only four-wheel-drive manufacturer in the European market. Even in the UK, Land Rover was only taking 50% of the available sales instead of the 95% it had been used to in the fifties and sixties. Rest of world sales still represented a healthy 15%. The government still had a great deal of faith in a company that, after the Land Rover, had introduced a second and completely unexpected success in the Range Rover (see pages 86-99). Downing Street sanctioned a £280 million modernisation programme for the factories which quickly paid a dividend. Initially, the Range Rover production facility was extended by a half again and the money also helped the Land Rover which by 1976, had become a millionaire in production terms.

The seventies saw the British motor industry heading for disaster. At the same time Rover was conducting the first crash tests on its vehicles. In the picture you can see the strengthened door hinges on the Series III

In 1979, Margaret Thatcher's conservative party came into power after a number of years of Labour government. At this time, there were over 165,000 people working for British Leyland. As part of the Conservatives' radical approach to economic policy, the highest priority was attached to privatisation of state industry. Key industries such as British Telecom were privatised and the motor industry was taken out of public ownership and reprivatised in 1986. Today, the marques Mini, MG and Land Rover are part of Rover group which also has the rights to other legendary names such as Austin Healey, Riley and Vanden Plas. In 1988, Rover was sold to British Aerospace. At this time, half of the company was known as Austin Rover, and the other half was Land Rover, but in September 1989, the decision was taken just to use the name Rover. In 1990 Honda entered the fray with 20% as a cross-shareholding agreement and offered a considerable amount of technical support, admittedly not to Land Rover.

Honda was keen to become more involved with Rover Group – a step the management of the day was keen to encourage – but BMW stepped in and bought the company (much to the surprise even of insiders) for £800m.

ROVER LAND ROVER MINI MG

The quartet of current Rover marques: Mini stands on its own and the MG badge (standing for Morris Garages from 1923 to 1976) was brought back to life

The introduction of the Series III saw the instruments now in a binnacle behind the steering wheel and new seats

Down under, people were missing the old-style radiator grille which doubled as a barbecue. The improved technology also improved the Land Rover's off-road ability

BMW did not get Jaguar and Daimler however, as they had been sold by Leyland in 1984 and subsequently taken over by Ford in 1989 for £2bn. In the maze of corporate identities that were being bandied around, Land Rover's brand image was and still is the jewel in the Rover crown. Whatever else may have happened, Land Rover would have survived any crisis – the marque was too strong.

Meanwhile back to 1971. September 1971 saw the launch of the Series III. Once again, the Land Rover enjoyed a make-over and very little of consequence had been done to change things. The interior saw the most revisions however, including a new steering wheel, and a completely revised dashboard with the instruments placed directly in front of the driver and not as before in the middle, representing the thinking of a more modern age. Among the most important technical improvements was a fully-synchronised gearbox, more powerful brakes – on the 109 there was even servo-assistance. The larger clutch which had thus far only being employed in the diesel and six cylinder variants, was now made standard equipment.

One of the more controversial points of detail on the new Series III was the plastic front grille as the metal version had done its time. Apparently the Australians were not amused by the replacement as it was hardly suitable for its traditional task as base for the barbie. The plastic grille stayed and apart from introducing an optional overdrive in

Typical Land Rover features: The bonnet-mounted spare wheel, the bumper-mounted winch, PTOs were now only offered on special order

LAND-ROVER

The V8 gave the Land Rover power like never before

Power from eight pots enabled the Land Rover to take on some new tasks including this snow plough in the Swiss Alps

response to the 1974 oil crisis, very little was done to the Land Rover until the end of the seventies. Throughout these crisis years, Land Rover sales remained a reliable backbone for the company, which brought turnover with virtually no investment. Sales remained constant around the 50,000 mark.

1979 also saw the introduction of the legendary 3.5 litre V8 Range Rover engine into the Land Rover, again only in the long wheel base variant. With a slightly reduced power output compared to the Range Rover, the Stage I Land Rover replaced the older six cylinder variant. To guarantee enough space for the V8 under the bonnet and fit the inlet restrictors for the carburettors, the grille had to be moved forward and the bonnet raised slightly. Today, its 91bhp at 3500rpm and torque figure of 166Nm at 2000rpm seem rather meagre when compared to the current 1998 Defender with its direct injection 2.5 litre diesel putting out 113bhp and powerful torque figures of 265Nm at 1800rpm. It must be added though, that the V8 implant for the Land Rover was detuned to the extent that its

In 1980 Mr.Land Rover himself, Tom Barton retired. True to form, he stayed on for a further three years as technical consultant advising companies all over the world

Land Rover as mule: A few changes had massively increased the carrying capacity of the Pick-Up

power output was a third less than in the contemporary Range Rover. In 1979 journalists viewed this very differently, writing about the first Land Rover that also had an acceptable turn of speed on the road (it could reach 80mph). The V8 engine also introduced other new technology to the Land Rover, namely the permanent four-wheel-drive system which had been introduced in the Range Rover in 1970. Interestingly, a limited number of 88inch V8s was also built for the West Indian market. Apparently, they were a wild drive!

As the seventies drifted into the eighties, the petrol and diesel engines were finally revised and the interior underwent another change. Additonally, the model range was widened to include a further pick-up variant which, because of its increased load tray was christened the High-Capacity Pick-Up. Without altering the wheelbase, the Land Rover engineers had been able to increase the load space by 20% and the total load volume (with a covered body) by 40%. Achieving this had been possible with some extraordinarily simple tricks. The inner wings behind the rear wheels were left exposed instead of being hidden behind the usual body panel which would normally run from the front to the back either side of the load tray. The rear tailgate was also extended across the full width of the load tray. This Pick-Up could carry 1080 kgs (Its GVW was 3.05 tons thanks to a strengthened chassis).

The Land Rover was capable of putting in a class performance with three axles. This is a special model called the Perentie with a V8 engine built for the Australian Army by Jaguar Rover Australia. According to Land Rover the original prototype was built by Bristol based firm SMC Engineering. New side mirrors and wing spats became standard features in 1983

The Australian Army had used Land Rovers for many years and some six cylinder Series II vehicles had seen active service in Vietnam. In 1982, the Australian Government issued a requirement for two types of vehicle – a medium duty 1 tonne and a heavy duty 1.5 tonne. The requirement called "Project Perentie" after a large Australian lizard specified significant Australian content, preferably being manufactured on the continent itself and involving Australian industry in offset agreements. The contract was to be decided by competitive trials. Following initial evaluation, Toyota, Jeep, Mercedes and Land Rover were selected for the trials phase with only Land Rover (with the 6x6) and Mercedes (with the Unimog) competing for the heavy vehicle. Following the trials, Land Rover was declared the winner and following successful commercial negotiations the contract was awarded.

Engineering and development was done in Australia by Jaguar-Rover Australia (JRA) by a team under the direction of Ray Hapgood.

The medium duty vehicle was based on the 110 but had little commonality with its British counterpart. The chassis was extensively modified to allow for stowage of the spare wheel between the side members at the rear in the space usually occupied by the fuel tank which was replaced by the 15 gallon side tank – normally available as an option! To give the ultimate in corrosion resistance, the chassis was also hot dip galvanised. The body was similar to a standard 110, but every opportunity was taken to provide storage locker space. The big change was in the driveline. The Land Rover engine was replaced by an Isuzu 3.9 litre direct injection unit which had already been offered in civilian 110s in Australia. This unit produced very high torque at low revs and had been marketed, with typical Australian candour, as having "More grunt than a yard full of pigs!" Thus equipped, the 110 was quite capable of completing a military trials course fully laden with a trailer in high range!

The 6x6 was even more remarkable. The chassis was totally new, although some 110 components were used. It was a space frame construction and, like the 110, was hot dip galvanised. The driveline featured the turbo-

charged version of the Isuzu engine. The drive to the third axle was achieved by using the power take-off facility on the venerable LR95 integrated gear and transfer box originally designed for the Range Rover, but by now superseded. To provide commonality, the 110 also used this four-speed unit. Interlocks only allowed six-wheel-drive to be engaged in low-range and it shared a common vacuum system with the diff-lock arrangement.

An ingenious system enabled for compensation between the two rear axles to

maximise ground contact and to ensure that no one axle would ever bear the entire payload. Axles themselves were wide track versions of the GKN Salisbury heavy duty axle although the 110 had a locally modified Land Rover axle.

The cab was designed to accommodate three fully-equipped soldiers and although of Land Rover appearance was wider and longer and constructed of aluminium over a steel space frame. The bonnet was a fiberglass moulding. The rear body was locally made and designed to be stripped to act as a flat bed truck. Seats were foldable and designed to be stored on the vehicle under the load tray. Ambulance and other specialist bodies were also produced including an open version for the Australian SAS.

The vehicles were very popular in service and a production facility was recently re-opened to build a follow-on contract. The 6x6 was marketed as a civilian vehicle and achieved some success in Australia.

Without a doubt, this model was one of the most interesting Land Rover variants. The concept is still alive today.

Solihull Crates

Why the Land Rover was sent in pieces across Europe

Minerva, the Roman Goddess of the craftsman, wisdom and the arts adorned the firm logo

Nobody would have guessed that this was on the cards. The Land Rover, which had originally been planned as nothing more than a stop-gap measure had become the best-selling vehicle in its class barely a few years after its introduction. The ever-increasing sales figures not only made the books look good, they also caused Land Rover something of a problem, namely Solihull could not match demand with the production capacity they had.

One early way of getting round the lack of production capacity and still sell Land Rovers came about in 1952 via the Belgian Minerva company. The Antwerp-based Minerva company, which had been founded in 1904, had been Belgian's largest car manufacturer and making a name for itself with some exquisite luxury- and sports cars powered by four and six cylinder pushrod engines. Financial difficulties caused Minerva to close down its operation in the thirties. Despite a merger with the Belgian luxury car manufacturer Imperia in 1936, the last Minerva rolled off the production line in 1939.

After the war, Minerva dared to make a comeback. As constructing their own vehicles was considered to be too expensive and therefore risky, Minerva decided its fortune lay in licence production. Consequently, Minerva boss van Roggen was invited to Solihull in the spring of 1951 after he had already been to talk to Willys in the United States to see if there was an opportunity going begging there; it was Rover however, which set the pace. By the end of the year, the two companies had agreed to enter into a four year contract for licensed production. On the back of the contract, the Belgian Army ordered 10,000 vehicles. The Belgians had already ordered ten Land Rovers in 1948 for their military and appeared to be more than satisfied.

Left-hand drive 80 inch chassis, two litre engines, gearbox and axles were supplied from May 1952 as CKD (Completely Knocked Down)* kits from Solihull and delivered to

**CKD kits were a very good way of avoiding the high import tax in some of Land Rover's export markets (to protect national economies).*

Early Minervas retained the Land Rover logo behind the rather brutal protective grille

the Minerva factory, where they were assembled and given a steel body. Later, even the chassis were manufactured in Belgium.

The Minerva models are easily recognised by their modified body, squared-off front wings, flattened main headlights, a modified front grille and an exhaust which came out at the front underneath the driver's door. Inside, there was no third front seat – in its place there was a tool box. The spare wheel was positioned at the back on the right, to the left was a spare jerry can holder and the tailgate was hinged to the right. A limited number of civilian models produced from October 1953 can be identified by a third front seat. Additionally, they were the only model to have exterior door handles and a rear tailgate. Just like their English counterparts, they could be ordered with a range of accessories and consequently tended to be better equipped. In 1954 and at the same time as Rover introduced its own vehicle, the longer 86inch appeared of which only 1,100 were built until the Minerva contract ran out. In total, Minerva had constructed approximately 18,000 Land Rovers. It is impossible to be more precise because a number of the CKD chassis numbers are in Rover's own production records.

Just weeks after the Land Rover production contract had finished, Minerva brought out its own off-road vehicle which it had developed itself. This was the C/M 20 (the C and M stood for Civilian and Military) and was a monocoque with the engine mounted in a subframe and came in two wheelbase lengths. Its climbing ability was 70%. Nevertheless, only a few were ever built as Minerva went bankrupt in 1956 and was broken up in 1958.

Rover had a second licencing opportunity in 1952. An almost forgotten German utility vehicle manufacturer based in the Hamburg suburb of Harburg heard of a tender being put out by the German Border Police. The relatively young German Federal Republic required a six-seater patrol car that was capable of being driven off-road on border patrol duty. Naturally, the Land Rover had been considered, but people had been put off by the extremely long delivery times. Consequently, the authorities started approaching German companies. For a while they were unsuccessful in finding a manufacturer able to come up with the goods. At that time in Germany, no-one was building a suitable car as the production of military specification vehicles was not allowed. As well as this, it did seem that the anticipated production run of only 250 vehicles would render any new design of vehicle financially unviable and therefore unprofitable. It fell to Tempo company boss Oscar Vidal to come up with a winning solution that would satisfy all parties. If they could not design and build their own vehicle, why not simply build the highly suitable Land Rover under licence? Measured against more than 20,000 Land Rovers manufactured in Solihull alone in 1952, the 250 Tempo Land Rovers were a drop in the ocean. Nevertheless, the men from Rover must have liked what they saw in Hamburg as, after a further visit by Spencer Wilks to the Hamburg plant, the contract was drawn up and signed between the two firms. It has been suggested that Vidal's contacts with the British occupying forces in Germany possibly had something to do with the signing of the deal. It was agreed that production of a limited series of the run out 80inch model would take place along the lines of the Minerva model. All the components were delivered by Rover in crates to Hamburg while the chassis and the body were built to German Border Police specifications by another Hamburg-based coachbuilder called Herbert Vidal & Co. – the name gives it away. It belonged to Oscar's brother.

Made in Hamburg: The Tempo Landy

The first Series of Tempo vehicles can be identified by their high bodysides. The exhaust comes out on the left rear side of the vehicle

The twin rear-mounted tanks under the rear bench were linked via a hose, the rear screen was made out of glass and the BR acronym stood for Britische Zone Rheinland. Finally the rear panel was completely sealed. It did not open

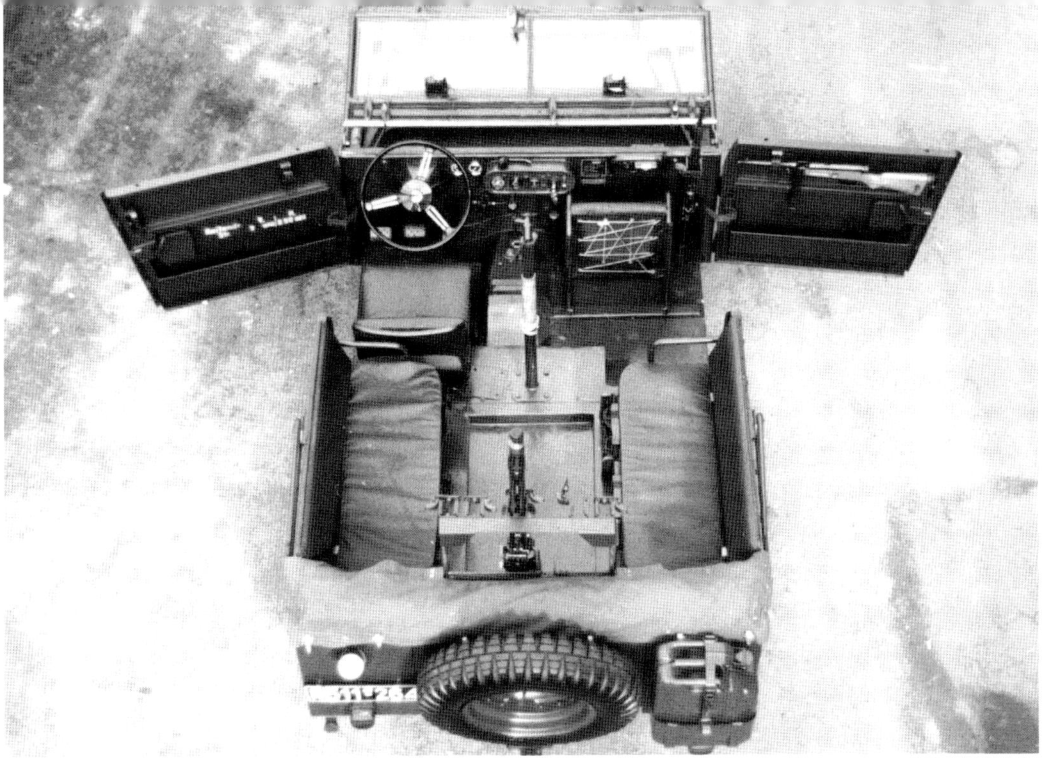

This is what the German Federal Border Police wanted. Six seats, four-wheel-drive and weapons racks

After a pre-production model had been brought over to Solihull for assessment and had been given the go-ahead for production without any changes being suggested, the first deliveries were made of German-built Land Rovers to the German Border Police in April 1953. Although a cursory glance at the Tempo would, just like the Minerva, confirm that this was indeed a true Land Rover, just like its Belgian cousin, the Tempo differed from the original in a number of ways. The most significant identifying feature of what was known internally in the Tempo factory as the Series I was its all-steel body with high body sides and additional storage compartments in the front wings.

By August 1953, 100 vehicles had been produced and then production was re-jigged for the longer 86inch model. For this revision, the body was also now made like a Land Rover from aluminium. The new version was also equipped with a deeper dashboard making the storage of radios and other equipment that much easier. The rest of the 250 vehicle order was then completed by 1955. Afterwards, Tempo decided to stay loyal to the Land Rover and continued to sell Solihull-built vehicles until 1965. The Tempos also remained in service with ordinary police units until they were all pensioned off in the sixties. Today, they are among the rarest of all Land Rover variants of and the Series I Tempo is particularly popular with enthusiasts.

Unique to Tempo: Storage boxes in the front wings, a bonnet-mounted stowage box for the door windows, a lockable cubby box under the passenger's seat instead of the standard fuel tank

A Santana deluxe model with chrome and tropical roof. In design terms the Spanish were actually further down the road than the British: This IIA has already got the headlight fit of the British Series III. Additionally a single panel front windscreen was available from 1972 onwards. Note too how the radiator grille has been brought right forward – a move which only happened towards the end of the seventies in Britain

Belgium and Germany weren't the only countries to produce the Land Rover under licence. In 1955, Rover entered into a partnership that would prove to be one of the most profitable and longest-lasting of its kind. Its gestation was however, unusual to say the least.

In the Spring of 1954, the Spanish Ministry for Industry embarked upon a programme of revitalising industry in the blackspot province of Jaen in northern Andalusia. Although lacking in good ideas itself, the ministry set up a competition to encourage investment into the area. One participant was a young industrialist called Alfredo Jiménez Cassina. His plan was to construct a factory for agricultural machinery right in the middle of the world's largest olive tree plantation – not surprisingly, he got government approval and sponsorship. Armed with his state subsidy, Cassina bought a piece of land together with some friends in Linares which they christened "Santa Ana" which was then shortened to Santana. The official company title was "Metalurgica de Santa Ana SA" (the SA stands for Limited Company in Spanish). Nevertheless, the Santana name would establish itself in the nation's consciousness. In 1955 however, as the first factory buildings were being erected, most were not quite sure just what exactly was going to be built in the factory. Of one thing they were certain; it had to be a universal all-purpose vehicle. It was only a matter of time before the young Spanish businessman began to consider Land Rover.

When Cassina contacted Solihull, the Rover team was neither surprised nor particularly enthused by the idea of licence production as they had amassed a number of years of experience in this area. Nevertheless, things did not turn out exactly in the way Solihull planned. The Spanish dictator Franco didn't just want cars. He insisted that most of the components be produced in Spain. In particular, according to the conditions set out, at least 70% of the vehicle had to be manufactured in Spain. Amazingly, Wilks

Spanish craftsmen were also fans. This 109 Pick-up with hard top and tropical roof was run by an electrical repair company on the island of Majorca

said yes to this. It was to be four years before the first Spanish-built Land Rover left the factory at Linares however. In 1959, Solihull was already building the Series II, so no Series Is were built in Andalusia. With some pride, Santana launched its first off-road vehicle in both short- and long-wheel base form (88- and 109inch). A diesel engine was offered as an alternative to the petrol variant and all models corresponded to the British equivalent from Solihull. Additionally, Santana produced its own engines based on Rover powerplants right across the model range and these were even delivered to Solihull when the UK factory suffered a shortfall. In Spain, the Land Rover success story was repeated, with economic success matching the vehicle's reputation for robustness and sheer popularity – a success story which had thus far been denied the Landy as a result of the high import taxation. Slowly but surely, the Santana plant began to increase production levels. Rover itself already had a presence in Spain grabbing 23% of the market and was now taking advantage of the increase in vehicle sales. Wisely thinking ahead, Santana invested its profits in the construction of new production facilities and buildings.

Now that the factory was up and running with more production capacity than ever before, Santana began to diversify taking on licence production of Citroën gearboxes as well as Commer Transporter trucks. Land Rover remained the firm's number one product however. Over the years, a number of very diverse models appeared, some of which were recognisably based on Land Rovers, others which were less so.

Santana's best customer was the Spanish Army. In 1967, the company built the 1300 which was related to Land Rover's Forward

A rare sight: A Series III in Seville with a glass fibre roof, one piece front windscreen and three windscreen wipers. It is very rare to see much older examples of Land Rover Santanas in Spain. The younger examples are more common

The Santana Lightweight – here with a raised snorkel for wading – had a very different-looking front end from its British counterpart

Control (see pages 57-61). The Ligero from 1980 was a particularly original design with a dismountable aluminium body, thus converting the vehicle into an airy, fun off-roader with limited utility value. The Ligero sold like hot cakes. Land Rover was so pleased with its Spanish subsidiary that from 1981 it was allowed to trade under the name Land Rover Santana. However this cosiness could not prevent dark clouds from the far east upsetting the partners somewhat. Some at Santana felt (unjustifiably in truth) that the then current Land Rover model range was no longer up-to-date and looked instead to the trendier offerings coming from Japan. Land Rover claims that the dispute had more to do with disagreements over marketing and in particular to countries in South America and Iran. This culminated in Santana entering into a licence agreement with Suzuki. In 1990 and five years after they had started building the Suzuki SJ, Rover decided to end their agreement with Santana. The main factor behind this decision was however not the fact that Japanese competitors were being built next door to Land Rover's own production line, but had much more to do with the fact that Santana was making huge losses at the time.

Santana had made the mistake of misjudging the potential sales for the Suzuki and kept too many people on in the factory. The 800 employees who became superfluous after the Citroën contract ended were kept on at the plant. Nevertheless, the Spanish continued to produce the Land Rover under

It's all a matter of taste: This is the civilian version of the Spanish military Lightweight known as the Ligero, complete with square headlights

the model designations 2500 and 3500 (the latter being the Spanish version of the Forward Control with a six cylinder engine) until 1994. In total, Santana built 287,067 Land Rovers of which 87,739 were 88s, 143,779 109s and 44,355 CKD kits and other special models which were assembled in overseas plants.

Rover's decision to pull out hurt Santana badly; short-sighted management had finally delivered the coup de grâce despite improving Suzuki sales to this one time prestigious company. By 1993, nobody was following the share value of the company any more and this encouraged Suzuki to transfer its capital majority of 83% to an Andalusian state holding company. In 1994, the production line at Linares stopped temporarily as the company was declared insolvent. In 1996 the company managed to turn over a small profit. Since April 1997, the company has been owned by the state.

In the year 2000, the licence agreement with the Japanese runs out. The plan is to build a new European off-road vehicle with Ford and Peugeot engines with an official launch date in 1998. Whether it will be as successful as the Land Rover remains to be seen.

1998 sees the assembly of CKD Land Rover Defenders in the Brazilian state of Sao Paulo. Backed by a local investment of £90 million and creating 800 jobs, the new plant is to build Defender 90, 110 and 130 for the local market. Initial production is to be around 2,000 vehicles a year.

This is what the last Santanas looked like. They differed in a number of details from Solihull vehicles. The four-door version had twin alpine roof lights and again triple windscreen-wipers. Even the indicators are different

On the left is an 88 Super with additional roof ventilation. Santanas generally suffer hard lives. Dents are the norm

A Class Act through the Mud

Nobody does it better: Range Rover

It is rare for a car to have an impact like the Range Rover had when it was launched in 1970

History never repeats itself – or does it? After going from nowhere to world number one with the Land Rover, Rover was able to repeat the feat 22 years later. The Range Rover "made by Land Rover" rapidly became the epitome of luxury four-wheel-drive and has remained so to this day. 24 years of production puts even the Series vehicles in the shade. The current 90/110 Defender series holds the current record (which is increasing as we speak) as it has been built for 14 years. Even after the launch of the new Range Rover in the autumn of 1994, the old shape, now named Classic, continued in production until February 1996.

This chapter of the Land Rover story has its origins in the mid sixties. The idea for a vehicle like this is however, considerably older. Maurice Wilks, the force behind the legendary Land Rover had the idea for a new vehicle which was able to venture off-road but which also had good road manners. A name had already been found for this new vehicle: Road Rover which linked rather nicely to Land Rover. Under Wilks' overall leadership and with Gordon Bashford as Chief Designer, a prototype based around the Rover P4 and equipped with a two litre engine was quickly put together acquiring the nickname "greenhouse". However, it was planned to have rear-wheel-drive only. Indeed the design which took many cues from the Land Rover, was too severe and ascetic even for the fifties. Most importantly however, the project died a death because it was estimated that it would cost too much to build.

Towards the end of the fifties, a second Road Rover prototype was developed which was much more similar to an ordinary car. Designed in the style of an American

Solihull was already thinking about a mixture of Rover and Land Rover in the early fifties. The Road Rover never made it beyond the drawing board

shooting brake (two doors, 4.16 metres long) the vehicle which, like the Land Rover was to have been constructed largely from aluminium panels, might even have been a success in the market place. However this project was cancelled as well in 1958 as the Rover board decided to concentrate exclusively on the Rover 2000 car. In the mid sixties, Spen King took up his uncle Maurice Wilks' ideas again. King is similar to Barton and Bashford in that he is something of a cornerstone in the Land Rover story. Initially, he was more involved in the gas turbine engine development programme, moving later to cars where he was responsible for the development of the above-mentioned Rover 2000.

Together with Bashford, King worked on the first sketches of a prototype which already had many of the styling cues of the

Styled along the lines of an American shooting brake, the second Road Rover study was cancelled towards the end of the fifties

As the luxury off-roader started to take shape, the old Road Rover name was still being used

production version in 1970. The concept was based around the idea of an off-road vehicle which would be as comfortable to drive as a car. Consequently, King felt the departure from leaf springs was imperative, suspension had to be coil springs and the most modern shock absorber technology available. Amazingly, King took a Rover 2000 off-road to gain some first impressions and the results were encouraging. In getting the optimum amount of suspension and shock absorption, more work was carried out than was otherwise usual at this time. The rear axle even received a self-levelling device manufactured by Boge and which had been discovered by King and Bashford at the Frankfurt Motor Show in 1967 where it was displayed on a Mercedes (Boge Hydromat).

One of the more colourful background figures involved in the extensive testing that the Range Rover underwent in the sixties was Major Benjamin Hervey-Bathhurst OBE. 'The Major' as he was universally known, owned Eastnor Castle on the English/Welsh borders and after an illustrious career in the British Armed Forces in World War II, was one of the first people to buy a Land Rover. Because of his interest in the vehicles, he decided to contact the company and offer them the use of his estate as a proving ground. It was partly as a result of the Major's comments regarding the then revolutionary Ford Bronco, that Tom Barton gave the Range Rover project team the go-ahead. Sadly the Major died in 1997. Land Rover acknowledged the debt owed by the company to this remarkable man in a moving obituary, "Along with the Eastnor estate, Ben Hervey-Bathurst became a cornerstone of our marketing and demonstration philosophy, really enhancing the brand."

Originally the Range Rover was to be propelled by Rover's old three litre six cylinder engine from the P5. It can be said with some certainty that had the Range Rover been powered by this engine, it would not have been quite the success story it turned out to be. The Range Rover's success was in no small part due to its now famous 3.5 litre V8 engine which originally came from America. Once again chance had played a role in events.

Were one to describe this engine as a marine powerplant, it wouldn't be too far from the truth, for when Rover's Managing Director William Martin-Hurst saw the engine for the first time during a business trip to the US, it was actually being prepared to be fitted into a speedboat. Martin-Hurst was intending to sell the American yacht builder Mercury the Rover gas-turbine when he noticed an engine that had been pulled only the day before out of a Buick Skylark. This amazing, extremely compact aluminium V8 impressed him straight away. (After all it was barely longer than Rover's standard four cylinder engine and not much heavier.) Martin-Hurst was also surprised to learn that General Motors had dropped the engine from its powerplant programme. All that was required was a bit of negotiating finery before GM gave Rover the go-ahead and sold the rights. Initially, the GM people did not take Rover seriously, not understanding why the Brits could possibly want a motor that had already been removed from their development programme. This is how a marine engine became the heart that was inextricably linked with the success of the Range Rover. As well as in the Range Rover and Land Rover, the 3.5 litre V8 was also conscripted for service in the P5, its successor the Rover 3500, (P6) and even in the Morgan and TVR sports cars. The latter two took the engine in the same way as Land Rover had and still do today albeit in much modified form. Meanwhile, a prototype of the new vehicle had been put together which had been named simply the Land Rover 100inch station wagon because the wheelbase after the first sketches had come out at 99.9 inches. The chassis was new, but the axles came from Land Rover. Literature written about the Range Rover often says that it was equipped from the outset with permanent four wheel drive. Spen King tells a different story; Originally a standard, part-time four-wheel-drive system was planned for the Range Rover, but because the propshaft and the differential were too weak to cope with the V8's power in two-wheel-drive mode, it was decided to go for the permanent option. Rover had already conducted some experiments along these lines in the fifities, in particular with the T3 (see page 42) which also had been developed by King and Bashford. Today, it is all too easily forgotten that the Range Rover was the first production car to have permanent four-wheel-drive and to make it acceptable to the driving public. A luxury sports car produced by the English firm Jensen with a similar set-up had appeared four years previously. In the case of the Jensen however, the engine power was not divided equally between all four wheels, after all, the Jensen FF was designed for the road. Only a few Jensen FFs were ever built.

Permanent four-wheel-drive with a locking centre differential immediately did away with the disadvantages of the part-time four-wheel-drive system. First of all, it didn't have to be changed into four-wheel-drive mode as it was permanently so. Thanks to the centre differential, you also avoided the problem of transmission wind-up on the road because of the differing revolution speeds of the two axles. Off-road, the centre differential could be locked up so that there was always engine power going to the second set of wheels even

Under the body the Range Rover hid a whole range of technical innovations which were ground breaking at the time. Permanent four-wheel-drive with a lockable centre differential, disc brakes front and rear and a self-levelling system on the rear axle

if one or more was spinning. Long travel coils was also the key. Centre differentials only work if all wheels are in contact with the ground as there is no axle differential. King was also aware of the advantages of permanent four-wheel-drive. King enjoys reminiscing about life in the sixties as an engineer. Speaking in his beautiful house in Leamington Spa, he has said that life as an engineer then cannot be compared to how things are done today, the individual had much more room to manoeuvre and there were fewer bureaucratic obstacles. "Compared to today, we had no development costs with the Range Rover and the Land Rover in 1948. We just did what we thought was right."

A reduction gearbox naturally belongs alongside a permanent four-wheel-drive with a lockable centre differential; After all, the Range Rover was supposed to be able to take the roughest terrain in its stride. Generally speaking, a reduction gearbox brings more engine power to the wheels and also gives better engine braking when descending sharp inclines. The LR95 reduction gearbox also pioneered the use of an integrated case.

In one sense however, the origins of the Range Rover were a result of modern thinking. At the same time, Graham Bannock, who was working in Rover's marketing department, had noticed a new trend in the four-wheel-drive vehicle market in the USA. There, the move towards a larger leisure-oriented off-road vehicle was becoming increasingly more obvious. At the outset of the sixties, it had been the arrival of the Jeep Wagoneer which had started this trend. Bannock suggested to the Rover management that the company build something similar. According to Spen King however, it wasn't this analysis which had given birth to his project. In 1967, the first prototype began road-testing and slowly the new vehicle began to come together. Bannock had meanwhile ordered a survey done of potential buyer interest in the new vehicle. According to the original brief, the vehicle would "offer safe, fast and relaxing travel on motorways, on the road, across fields, sand, snow or ice. It is easy to drive and handle, will tow boats or caravans, carry milk churns, be used as a camping vehicle or you can use it to go shopping or to go to the opera." So, it would seem that it was indeed Bannock who initiated the long-term journalistic running gag about the Range Rover being the off-roader you could take to the opera. Those people who had been questioned in the survey were largely positive about the vehicle, particularly when they learnt that it was a Rover.

After technical director Peter Wilks (just like Spen King, he was a cousin of Maurice and Spencer Wilks) had driven the prototype and given it his seal of approval, the decision was made to go ahead and build the car. The styling stayed pretty much as it had been on the prototype just as King and Bashford had intended it to look. Designer David Bache made some decisive changes including the castellations on the bonnet – a feature that was subsequently carried over to the 1994 'new' shape Range Rover and even the Freelander. On the first prototype, Bache had kept the name "Road Rover" across the bonnet. Up until this point, the design of the interior had not even been considered. This was to be left to Tony Poole and his team.

The original plan to offer the vehicle with a cheaper specification including the smaller Land Rover engine was dropped. The team had already made two decisive errors (admittedly more obvious with the benefit of hindsight) which fortunately did not impact heavily on sales of the vehicle. First of all there was no four-door version because with the resources available, the team could not compromise the integrity of the two-door body structure and in any case it was thought at the time that a leisure vehicle in this

Viewed today, the interior of the first Range Rover is hardly what one would call luxurious. Spartan would be a better word

Dissimilar brothers: A Series III next to the Range Rover which was barely altered in its first year of production. From 1977 the side mirrors were moved to the A-post

category would not need four doors. Secondly, it was not offered with an automatic gearbox. What was offered was a four-speed manual with what the renowned Land Rover historian James Taylor in one of his many books on the subject refers to as "the longest throw in history". In 1968, the name of this new vehicle was finallly decided on. Tony Poole had suggested Land Rover Ranger and this eventually became "Range Rover". Although the attempt was made to give the interior a luxurious air, placing the instrumentation seemingly at random all over the dashboard was felt by many to be a major shortcoming. One of the Range Rover's more novel features was the seatbelt integrated within the seat itself. This was a decision made after engineers discovered that locating the belt in the more usual place, ie the B-post, would have made it uncomfortable for passengers. In December 1969, the first three pre-production vehicles went down the newly-installed track at Solihull. In the meantime, Tom Barton had insisted that every Range Rover leave the factory with the words "made by Land Rover" on the back. In June 1970, the company's great hope for the future was presented to the press. The press was overwhelmed with enthusiasm. It seemed as if the world had been waiting for this car, which, with its V8 engine was good for over 95mph. The demand was so great that in the first few years of production, the company could not build enough of them. And all this for a vehicle which didn't have power-assisted steering on the first models. On the plus side, they did have disc brakes all round. Viewed with today's critical eyes, the interior can hardly be described as luxurious, in particular, the easily-torn pvc-covered seats also looked horrible. Electric windows and other toys were however, not on the cards. If you wanted to adjust the side mirrors, it meant getting out because they were located on the front wings (in line with contemporary practice).

And yet, at that time there was no competition for the Range Rover and its reception was incredibly positive everywhere.

101 Forerunner: In 1968 a new Forward Control prototype appeared

The German Frankfurter Allgemeine Zeitung reported in November 1972 that the Range Rover was equally at home whether on motorways, log roads or gravel tracks. The steep price came in for particular criticism, as did the vehicle's height above the ground and naturally those missing two rear doors. In 1970, the price in England when the vehicle was launched was £1,999. As well as this, the FAZ road tester and London Correspondent, Kurt B. Hopfinger complained about the way the foot pedals were weighted and about the sparse dealer/service network. Otherwise, the vehicle received praise everywhere for its performance and driving characteristics. It must have had something to do with the way individual components were matched up with each other. Other than that, the set-up was fairly conventional. The fuel consumption of 16mpg was considered to be quite normal. "It even has a beam axle up front". This was of course, three years before the oil crisis.

Publicity and reputation for the fledging came with the Darien Gap Expedition. Two Range Rovers were the first vehicles ever to travel north to south from Alaska in North America to Chigorodo in South America. The real challenge on this expedition however, was the crossing of the Darien Gap, a four hundred kilometre long swamp between Panama and Colombia. The famous Panamericana road was broken by this gap at the time. It took 99 days to cross this area. Without the help of a complete team of technical people and spare parts flown out specially from the UK – the differentials for example were not up to coping with the stress of the over-large swamp tyres with which the vehicles had been fitted – the expedition would not have happened at all. One particular irony was that the expedition was also completed by a used Land Rover that had been purchased in Panama. The British Trans-America Expedition required a good six months from December 3 1971 to June 10 1972 to complete the journey. However the two Range Rovers and the 88inch Land Rover were not the first vehicles to overcome the Darien Gap. Jeeps had been successful and two eccentric Englishmen had completed the journey in 1960 in 134 days – naturally in an 88inch Land Rover.

The fact that the Range Rover was so far ahead of its time probably guaranteed its

Darien Gap proved to be one of the most strenuous tasks to face any four-wheel-drive vehicle

long-term existence for its introduction coincided with the crisis that befell British Leyland at the same time. There was no money for further development of the car and this is one of the reasons why it stayed as it did for so long. Additionally, it wasn't even possible to extend the production facilities.

In 1973, production of the old Forward Control stopped. Its successor was the 101 known to the military as the "1-Tonne" utilising the V8 engine and permanent four-wheel-drive system from the Range Rover, but keeping the leaf springs from the normal control vehicles. In 1972, the vehicle was given its first airing and three years later it went into production albeit solely as a military vehicle. In 1978, production of the 101 was stopped and thus ended the era of the Forward Control Land Rover. Many 101s have since found their way into the hands of civilian owners after being demobbed by the British Army. Off-road it is virtually unstoppable, particularly when equipped with the Rubery-Owen powered axle trailer. This 6x6 team tends to make light work of most obstacles. Proof positive of this combination's sheer ability came in the crossing of the Sahara from Dakar to Suez with two so-equipped vehicles in the mid seventies. As with the Darien Gap expedition, this was an initiative of the British Armed Forces. The 101s required 100 days to travel the approximately 12 000 kilometre distance.

In the mid eighties, Land Rover tried to enter the market with another Forward Control. Llama was born out of a UK Ministry of Defence requirement for a vehicle to replace the 101 inch 1 tonne forward control to be known as "Truck Utility Heavy Duty". The vehicle had to carry a 2 tonne payload but only weigh 2 tonnes unladen, so that it could be suspended under the helicopters then in service. Contemporary 2 tonne vehicles such as the Renault TRM 2000 had a gross vehicle weight of 6.3 tonnes to achieve a 2.3 tonne payload!

The requirement also specified a diesel engine and stated that automatic transmission was preferred. The MoD did not want a bespoke vehicle and insisted that the candidates must have some „commercial" sales prospects to keep costs down. A forward control design was preferred and derivatives such as cargo vehicles, Fitted For Radio (FFR) and ambulances were envisaged. Other manufacturers approached included Stonefield, Volvo and Reynolds Boughton.

The military market was an important one for Land Rover and a team was formed to produce a vehicle to meet the MoD requirements.The first concept was effectively a 110 with minimum changes to accept a tilting cab and the necessary linkages for the gearbox and engine. The cab design was, based on glass fibre panels which could be produced on "soft" tooling

Left: The FC 101 was virtually unstoppable off-road when equipped with a powered trailer. Above: Although Land Rover's first forward control had not exactly been blessed with success, the company took the plunge a second time with the 101-a forward control based loosely on the engine and drivetrain of the Range Rover

Third and last attempt: Three years after the final 101 had rolled down the track, experiments were conducted with this prototype codenamed Llama. The driver's cab was made out of fibre glass and the design is closely reminiscent of Leyland trucks during the period

hung on a steel tube fabricated frame. The cab design and construction was undertaken by Motor Panels in Coventry who were also manufacturing Leyland truck cabs and the design naturally reflected those influences. The prototypes were equipped with the V8 engine and LT85 five speed gearbox. Studies were undertaken on alternative diesel engines of larger capacity such as the Perkins "Phaser" range, Cummins "B Series" and even the Isuzu unit being used in the Australian Army "Perentie" vehicles.

Though it had less power and torque than the proprietary 4.0 litre engines, the much lighter "Gemini" engine, which later became the Tdi, was also considered. Initially, Salisbury axles were used from the 110 heavy duty specification with thicker axle tubes, heavy duty front swivel pin housings and bearings. Problems with stability when loaded with an ammunition pallet, led to wider axles being used and the vehicle was shown at one military show with a wider rear track. A limited slip differential was fitted to the rear axle on later vehicles.

There were considerable problems with chassis durability when fully laden and stowing the 9.00x16 spare wheel proved difficult. Space requirements for the crew, weapons and equipment were difficult to meet in a vehicle of this size. However, despite some further work, the problems could not be resolved. Land Rover eventually stopped development of Llama and offered the Australian Land Rover 6x6 to meet the requirement, but this was turned down as being too complex.

Llama therefore passed into Land Rover legend. Perhaps a radical chassis re-design on the Australian Land Rover pattern would have resolved a lot of the problems of stability and spare wheel packaging but it was unlikely that the power train issues could ever have been solved. Some Llamas survive in private collections in the UK and with the Heritage Motor Centre.

Meanwhile on the Range Rover, there had been some progress. 1973 saw the introduction of power steering and a rear windscreen wiper. Additionally, the reaction to the oil crisis was to give the vehicle overdrive via Fairey. Otherwise the changes made were largely cosmetic including a move for the wing mirrors from their rather exposed outside wing position to the A posts.

For over ten years very little of note was changed on the Range Rover. Nevertheless, Range Rover sales figures increased. Today, this simply would not happen. With government money however, (see page 71) things began to change as the seventies changed into the eighties. The production facilities could finally be modernised; in 1981, eleven years after it was launched, the Range Rover received a set of rear doors and in the four-door version, electric windows. The Swiss Monteverdi company had built their own four-door version of which approximately fifty were manufactured before Solihull and their move was partly responsible for prompting Land Rover to produce their own four-door design.

Now things began to happen in rapid succession: An automatic gearbox was finally made available in 1982 which also coincided with the 100,000th Range Rover to be built. Initially, customers had the joy of changing gear via a three-speed Chrysler 'box, this was replaced however, with a four-speed unit from the German manufacturer ZF in 1985. 1983 saw the introduction of a five-speed manual gearbox. The introduction of the "Vogue" specification in 1984 marked the beginnings of the Range Rover's development as a true luxury liner. At the same time however, the number of standard features and optional extras increased dramatically as did the sales figures. In 1987, the first year in which more Range Rovers than Land Rovers were sold, the power output of the V8 engine was increased to 165bhp. In 1986, after having been in continuous production for 16 years, a second engine was introduced into the range; an oil burner. By this time, diesel engines had become so acceptable to the chattering classes that a home could be found for one even in the Range Rover. The hard facts of European market demands also meant that there had to be a diesel option for countries such as France, Italy and Spain. Obviously, there were considerable tax advantages in those same southern European markets too. The decision was made not to use a Rover engine though. Instead, the choice fell on the Italian VM company and their 2.4 litre unit putting out some 112bhp which, with a max torque figure of 183Nm promised acceptable on-road performance and much reduced fuel consumption. In rapid time, the diesel version had taken 40% of the sales.

In 1987 Range Rover was launched into the North American market through the company's own importer, Range Rover North America (RRNA) based in Lanham, Maryland. According to research conducted by the company prior to the launch, there was considerable awareness of the Land Rover name despite the fact that Land Rovers hadn't been sold in the USA since 1974. To increase brand awareness the Land Rover oval badge was introduced onto the front of the Range Rover grille and has since been carried over onto vehicles for all markets. From an initial network of just 60 dealers, the current number in the USA has grown to 115 including 65 Land Rover adventure centres.

Just before the Range Rover celebrated its 20th birthday, the capacity of that solid, old "marine engine" was increased to 3.9 litres with 188bhp. In 1992, the diesel variant returned to its Solihull roots by taking the Discovery's direct injection diesel under its bonnet. Before it reached the end of its own

1981 saw the introduction of the four-door version. The door handles had also been revised with this in mind

New standards were set in 1992 by the introduction of air suspension. The lower sit mode made loading and getting in and out much easier

road, Range Rover broke its own record books all over again. In 1989, the production reached 28,513 vehicles. This record was only broken by the new Range Rover in 1995 when 29,619 vehicles came off the track.

Meanwhile, the competition had not been idle. The extraordinary combination of engine and equipment that was Range Rover was no longer unique in the four-wheel drive sector. However, the engineers from Solihull had something else up their sleeves. In October 1992, air suspension and electronic traction control were introduced. Improved versions of both these systems were subsequently fitted into the new Range Rover. The system which today goes by the acronym EAS (electronic air suspension) has the following as its most important components: Four air springs, four height sensors, a pump, a ten litre air reservoir and a control unit. The Range Rover has five different ride heights. Three of those can be set manually by the driver including standard mode, off-road (four centimetres higher) and a park mode to facilitate getting in and out of the vehicle (six centimetres lower). The vehicle also automatically sits down by two centimetres if the vehicle is driven over 50mph for more than 30 seconds. Finally, the air suspension automatically lifts the vehicle by three centimetres if it has come into contact with the ground. The system also did away with the need for the self-levelling device.

As well as an extension of the wheelbase (the 108inch LSE was an option), 1992 also saw an increase in engine capacity and performance with the introduction of the 4.2 litre engine with 202bhp (its max torque was 340Nm at 3 250rpm). With its lengthened wheelbase, the Range Rover now measured 4,65 metres. The 20 centimetres growth could be found in the wheelbase (2.74m instead of 2.54m) and resulted in more legroom for the rear passengers. In the back it was almost palatial. It goes without saying that all manner of accroutements including a sliding sunroof, air conditioning, airbags and ABS were either standard in the Vogue or could be had as optional extras. A few months before the new shape Range Rover was launched, the two-door variant was finally allowed to die. Solihull delivered the last one to Portugal in January 1994.

Seven months later the new Range Rover was presented to the world's press in England. It is not often that a vehicle has been so keenly-anticipated by the motoring press as the new Range Rover was. Had the designers succeeded in not distancing themselves too far from the previous model as they had managed to achieve with every revision thus far? In short, no. The official line from Rover sounded rather different however. According to the press release at the official launch, the new Range Rover was immediately recognisable as a Range Rover

Facing off: The new Range Rover was presented to the world's press in September 1994. The massively-styled body took some getting used to

as its classic lines had been adhered to. Eight separate design elements were referred to which had been kept by the designers. For the man in the street however, this figure can be reduced to the bonnet castellations and the horizontal split tailgate. German journalists in particular were quick to criticise this solution because it makes loading and unloading more difficult. The British have always maintained that the rear tailgate is a perfect place to have a picnic. Actually the Brits are right and as well as that, extra long items can easily be transported by merely opening the rear screen – a solution that has been given the modern treatment in the Freelander (see pages 118-123).

Range Rover enthusiasts were initially shocked by the appearance of the new model – it wasn't for nothing that you could still buy the old shape (now re-badged as the Classic) – however after three years of production most have got used to the new model which is still the King of off-roaders.
Technically speaking, it was a completely new vehicle. Land Rover had invested £750m in project director John Hall's baby. The chassis too was a new design. The welded steel ladder-frame chassis is much stronger than the previous construction and is also designed to absorb crash damage more effectively. The beam axles – a must for any real off-roader – are both lighter and stronger. The air suspension system was

Control in comfort: The interior set class-leading standards in the luxury four-wheel-drive vehicle sector

further improved, only now you can actually drive in the sit down mode at up to 20mph. The extra 6.5 centimetres that you gain can make all the difference in some particularly low garages. At 1.85 metres, the new Range Rover is three centimetres taller than the older version. Lengthwise, the new vehicle also grew by six centimetres to 4.71 metres. As with all Land Rovers, the body skin is fabricated largely from aluminium. These include the front wings, the door skins and the lower half of the rear tailgate. Electronically too, the vehicle was a massive improvement over anything that has previously come out of Solihull. The BeCM

The 4.6 HSE is still the vehicle to be seen in, whether in town or up country

(Body Electronic Control Module) controlled no fewer than seven separate computers on board.

Just as expected, new standards were again set by the powerplants. The good old 3.5 litre V8 had now grown into a modern, electronically refined 4.6 litre V8 whose 224 horses could propel the Range Rover to a top speed of 140mph. Alternatively, there is an equally newly – developed 4 litre V8 with 190bhp. The launch price for the 4.6 HSE was £46,550 for which you get "the lot" in terms of specifications. The 4.0 litre version at £37,950 is rather more modest in its aspirations: You don't get the traction control, automatic gearbox, electric sunroof and the electrically operated seats: Additionally you have to make do with smaller rubber (235s instead of 255s). The third version on offer is particularly interesting as it includes a BMW turbodiesel. It is priced at the same level as the 4.0 litre version. A further option is an even more tightly specified variant (no climate control, different wheels, no cruise control and no leather interior) for £36,250.

Though radically restyled, the new Range Rover retains all the presence of its classic predecessor

The installation of the BMW engine had been agreed before the takeover. Previously employed in the 5 and 7 series BMWs, the engine puts out 136bhp. Thus equipped, the Range Rover is much more economical returning around 25mpg – a saving of approximately 50 % compared to the V8 engine.

Demand for the new Range Rover is uninterrupted. The top of the range 4.6 HSE continues to serve that well-heeled sector which the Range Rover has traditionally made its own. The vehicle remains the choice of the landowner, luxurious yes, outrageously extravagant, no – unlike some of the more showy limousines in this price class.

Customer surveys have revealed many owners to be quite artistic types, prepared to tolerate some of the Range Rover's foibles – which one hears about everytime owners are asked about their well-loved cars. As an aside, the original, classic Range Rover is the only car ever to have been exhibited in the Louvre.

If one believes the rumours, the future for Range Rover includes a BMW V12 engine. Were this to happen, then Range Rover would have made the leap ahead of the competition once again, just as in 1970.

RANGE ROVER
MJP 563

Instead of "110", people now started talking about the "One Ten". The windscreen is a one-piece affair. Note too, the lockable bonnet

government. For the 90, the diesel engine was also given a new lease of life and the enlarged 2.5 litre unit now pumped out 67bhp.

Although the new model put fresh impetus into the Land Rover, production dropped and it also began to lose sales. In 1980, the company had managed to sell over 50,000 Land Rovers. In 1986, this figure dropped dramatically to just over 20,000. The off-road market was no longer being fought over by just Jeep and Land Rover. A whole host of Japanese manufacturers were producing what appeared to be better product (if one believes that sales figures tell the whole story). The Range Rover was simply too expensive to be manufactured en masse. As already mentioned in the text, the Range Rover was actually performing better in the market place by the end of the eighties than the Land Rover. It was obvious to all that the Land Rover was living on borrowed time and that its reputation only counted for so much. Despite the diesel's performance hike, the engine was still lacking in power compared to what the competition was

Still looking good: The Land Rover 110 Station Wagon

Opposite: Alloy wheels and spats, metallic paint and a more luxurious interior than the Wilks brothers would ever have envisaged for the original Series I define the Defender 90 today. Off-road, it is still unbeatable

COUNTY
DEFENDER
Tdi

Visually just as it ever was, a 90 Pick-up at work. In a the background is the Station Wagon

The new Land Rover was also configured to take the V8 engine

offering and the V8-engined version could only be offered with the four-speed gearbox as the five-speeder was not capable of coping with the power. In Australia, which has always been one of Land Rover's strongest markets, the CKD Land Rovers were offered with a 3.9 litre Isuzu diesel engine which put out 96bhp. Incidentally, the same engine was also developed for the Perentie.

The new five-speed gearbox was finally made available in 1985 and was a boon particularly for the V8-engined variants. At the same time, the five-speeder became the standard 'box and the 90 was offered with the V8. In August that year the petrol four cylinder engine's capacity was increased in line with the diesel bringing it up to 83bhp and 180Nm.

It was the diesel which caused Land Rover most problems. After plans to build a V8 Turbodiesel as a joint operation with engine manufacturer Perkins came to nothing (the much rued Project Iceberg) it was decided to turbocharge Land Rover's own diesel. This was duly introduced in October 1986. It wasn't simply a matter of bolting a turbocharger on top of the old engine. A great deal of work had to be done so that the engine could stand the increase in power. This included the construction of a new cylinder block. However 85bhp and a maximum torque figure of 150Nm at 1800rpm weren't exactly figures that Land Rover could shout from the rooftops. In comparison, the V8 engine in the Land Rover went from strength to strength. In 1987 it was putting out 134bhp. New detail changes also included exterior door handles which unfortunately had a tendency to stick expecially in sandy climes. Up until this point, the Series III was still being produced, albeit in limited numbers.

As the eighties drew to a close, Land Rover began to respond to the Japanese challenge. the third model in the programme was launched in 1989. This was Discovery (see pages 112-117) One year later in order to bring some sense of corporate cohesion into play, the tradition of naming the vehicles after their wheelbases was dropped. From this point on, the 90/110/127 range was simply named "Defender". Allegedly the name came about in a sleazy Boston pub while Rover employees from both the USA and the UK were having a drinking session. According to Land Rover sources, the name Defender was already being used as a project

How to recognise a Defender: Black-painted bumpers, and new vehicle badging front and rear

Time slip: The current basic version of the Defender 90 has a lot in common with the vehicle launched in 1948: Green, short wheel base, canvas tonneau and a load tray. With some vehicle specifications, the front windscreen can still be folded down

The Landy is not just a utility vehicle, it is also fun

designation long before any pub talk. Concurrently, the name Land Rover was elevated to the status of a brand and the dash which had traditionally been written between the two words was dropped.

Despite this sales dropped to a new record low and only 17,950 Defenders were built in 1993. One would have to look back to 1953 to find similar figures. In 1993 power steering was made standard and a new, smaller steering wheel similar to the Discovery was introduced. Far more important than these changes however was the brand new engine that was taken from the Discovery. The direct injection turbo diesel (which was codenamed Gemini) was certainly state of the art when it was introduced and the performance figures of 113bhp and a maximum torque figure of 265Nm at 1800rpm were top of the class. The older four cylinder engine in both its petrol and diesel form was slowly withdrawn from the market. Today the Tdi is the only engine on offer. The V8 which has since been extended to 3.9 litres with 183bhp is retained for a number of other markets including the USA where it is available in the 90. In this market the NAS 90 with the V8 is aimed at

This is how the Americans like it: A North American Specification (NAS) Defender 90 softtop with the lifestyle trimmings and the military style two-piece doors

Easy going: The US version has an auto 'box and air con

the powerful short wheel base Jeep. The NAS could be ordered as a soft top and as a station wagon. Additionally a resin hardtop was also made available. Interestingly, it comes with the old military style doors with interior handles as these are the only doors which accept the separate window sections, and which, in order to emphasise the convertible feeling, can be left at home. The Defender 110 was also offered with the V8 engine and a similar specification (though not as a softtop!) This off-roader is also now offered with an automatic gearbox which is a first in the Defender as was the air conditioning which came as standard. In order to satisfy strict American crash laws, all US Defenders are fitted with a rollover safety cage. From 1993 onwards, all Defenders were equipped with disc brakes front and rear, a new smaller steering wheel and power-assisted steering. The Defender is also sent across the world as a CKD-kit. Although it doesn't go out to the 19 countries where Land Rovers used to be assembled under Land Rover management until 1978, the CKD set-up is still an important part of sales and distribution. There are even new factories being opened. The most recent is in South Africa where the Midrand headquarters was officially opened by the Queen in 1993. At the Rosslyn production plant there is very close contact to BMW; Munich's limousines are built there next to Land Rovers. There they haven't shied away either from giving the Defender a 2.8 litre BMW six cylinder with 193bhp. Land Rovers are also assembled in Australia, Kenya, Malaysia, Morocco, Zimbabwe and in Turkey.

Fifty years after it was born, the original Land Rover is fitter than ever and this is reflected in its sales figures. Last year 30,000 Defenders came off the production line – the last time that more than 30,000 had been built in any one year was 1985. It does appear that the decision to retain the Land Rover's distinctive face has been paid back in full. It is one of the last originals on the market in its sector. The decision to return to the German market with the Defender would also appear to have been spot on too. Previously, Rover Germany had decided

Bavarian muscle: In South Africa you can have the Defender with BMW's current 2.8 litre six cylinder powerplant

Understatement: The BMW-engined Defender is good for 113mph

against importing Defender because the limited sales would not have made it economically viable, much to the joy of a number of private direct importers. After the BMW takeover, interest in Land Rovers rose visibly. Currently, six separate Defender variants are on offer in Germany including the basic utility version with a canvas tonneau and a simple load tray. This represents the most original form of the Land Rover there is. The Defender 90 also comes as probably the world's shortest seven seater. As in the beginning, three passengers can be squeezed up front and four can be sat parallel to the direction of travel in the back. In the 110 wheelbase the Defender seats nine. Instead of three seats up front, there is a cubby box in between the front seats because ten-seater vehicles in Germany require a special driving licence. Additionally there is the High-Capacity Pick-up variant. The top model is the 130 Crew-Cab. This vehicle is unique in Germany with its standard canvas tonneau for the load tray and a four-door double cab. Its handling on road is very different from the 110. Thanks to its much longer wheelbase, it is more comfortable to drive, but the added weight means that the engine has to work harder.

The Defender, (that is to say the Land Rover) has been on sale since 1948 and 1.7 million of them have been built. Whether a new model Defender will be built remains to be seen. At the Detroit Motor Show,

Box like, practical, the best: Defender 90 soft top. Instrumentation follows the maxim Keep It Simple Stupid (KISS). The new steering wheel looks bulky but feels very comfortable

OK, I give up, I am the greatest (and the longest): Land Rover 130 Crew Cab

Mud and dirt are the best make up for any Land Rover. This should prove no obstacle

Playful antics: Whatever it is that makes off-roaders want to get their vehicles wet, remains a mystery to most. However 50 centimetres wading depth should be enough fun for most

Well-equipped for the next adventure: A V8-powered 110 complete with accessories in South Africa

There is only one Land Rover: A 110 explores the Scottish Highlands. Over 70% of all Land Rovers are still running. Today three-quarters of all vehicles built are exported. In 1990 the company was rewarded with a Queen's Award for Export

Dr. Walter Hasselkus announced that all Land Rover products would be the subject of a five year review. Traditionally, the word from Solihull is something along the lines of "We never discuss future product". Allegedly, a new Defender will appear sometime in the next century. What is to be expected from a Defender for the new millenium are airbags and ABS. For the 50th Anniversary of Land Rover, a limited edition Defender 90 along the lines of the NAS 90 and equipped with the silky smooth V8 married to an automatic gearbox will have no problem finding takers. For Land Rover fans one thing is certain, the next Defender has to remain true to the classic image of the Land Rover. Once again the miracle has to be worked that manages to combine progress with tradition.

On trial at the Gerotec proving grounds in Johannesburg. The Defender's climbing ability is rated more than 100%

Bold Responses

The empire strikes back: Discovery

New kid on the rock: Rover's third off-road vehicle made a mark for itself very quickly

All good things come in threes. The same is true for Land Rover. In 1989, that is 19 years after Land Rover's second product (not including numerous other projects), the Range Rover had been introduced, a new member was added to the family. This was the Discovery. Indeed the years 1948/1970/1989 could stand for the changing of the generations. In its way, the Discovery is a sort of half-way house between the rough and ready Defender and the luxurious Range Rover. The massive sales show the importance of a vehicle which had been designed to fill the one gap in the off-road market Land Rover wasn't addressing, not to mention the importance to the continued prosperity of the company. It is fair to ask the question: Where would

Land Rover be without the Discovery? In 1996 almost 70,000 came off the production lines. In sales terms, the Discovery has been by far the most successful production vehicle that Land Rover has ever built. Not even the Defender in its best years comes close.

Rover actually made the decision to build the Discovery in 1986. During the previous year various models had been designed to fill the obvious gap in the market. An answer to the Japanese challenge was already long overdue. The launch of two new Japanese models, the Isuzu Trooper and the Mitsubishi Shogun prompted Land Rover to redefine its target market. Range Rover was too expensive to compete with the newcomers, and the 90/110/127 range's product and design characteristics were too utilitarian. Research was commissioned in 1984 to estimate the potential and likely growth rates of the newly-defined "4x4 leisure vehicle sector". As a result, it was considered essential that Land Rover develop a product to compete in the new sector and the first task was to decide whether this should be on the Range Rover or the Land Rover platform. A study called Project Swift was started at the beginning of 1986 and concluded that Range Rover should be the platform and this evolved into Project Jay to begin concept development within a newly created project team.Tony Gilroy who had been Land Rover's Managing Director since 1983 pushed the project on. The development team led by Mike Donovan, who has since made the move on to Rolls Royce and then British Aerospace, had only three years to get the project into production.

Fortunately the work on the new direct injection turbo diesel engine that the Defender needed so urgently had already started. It was obvious to all concerned that this engine would be perfect for the new Discovery. Chassis and bodywork were based loosely on the Range Rover, with a common 100inch wheelbase. The design however, turned out to be particularly original suiting its projected place in the market place perfectly.

Ever since its inception, people have argued about whether the Discovery is a pretty swan or a pregnant duck. It is all too often measured against its classic sibling the Defender, a comparison which is perhaps slightly unfair. The motor press saw the Disco as an easy target, and classified the vehicle as belonging somewhere between the plains of Africa and the local supermarket car park. It isn't the tough, minimalist off-roader that is Defender, but it isn't as luxurious and expensive as the Range Rover.

The raised rear roof with the alpine lights in its sides and the relatively small front windscreen are reminiscent of the Defender. The two inward-facing rear seats in the rear positioned in the sides of the vehicle and facing in which make the Discovery a seven-seater hark back to older times. The interior was blessed with a radically new approach which is often and solely ascribed to the noted interior designer Terence Conran, but which in reality was largely the work of Land Rover's own styling department. After eight years and a mid-life update, it still feels modern and refreshing. The integrated headrests are unfortunately a little too low

The first Discovery's interior was much more user friendly than people were used to in the Defender

One year after launch the four-door version was also available

Off-road the Discovery is just as capable as its siblings. In recent years the Camel Trophy has used Discoverys

In order to discover adventure and freedom, one apparently has to drive a long way

and some of the instruments are not exactly ergonomically placed, but the numerous storage bins and nets that litter the interior of the Disco – the nickname wasn't long in coming – put most other people carriers and leisure vehicles in the shade. It is tedious to praise the all-round view and high seating position that is common to all off-roaders. It should be mentioned that the rear of the Discovery cabin is a good place to be – there is enough space for three and the headroom is amazing. Off-road, the Discovery makes no compromises. A ladder-frame chassis, beam axles and the body on top. The body is made of aluminium sheeting wrapped around a steel skeleton. The roof is made out of steel. The rear door is mounted on the right opening across the full width of the vehicle while also having the function of carrying the spare wheel. Permanent four wheel drive, a lockable centre differential and a reduction gearbox for off-road work are, along with disc brakes and power steering, self explanatory.

A significant factor in the Discovery's success is its diesel engine which, despite its power, is particularly frugal. Fuel consumption of around 29mpg is the norm. It is true to say that the 2.5 litre powerplant, complete with turbocharger, intercooling and direct injection offers everything that has become accepted as the norm in cars today:

113bhp at 4000rpm. The maximum torque figure of 265Nm at 1800rpm tells more of a story for it is perfect for off-road work. The engine was developed in conjunction with the Austrian firm AVL. The cylinder capacity of 2496cc reveals family ties to Land Rover's old 2.5 litre diesel and it is true that this engine was taken as the basis for subsequent development work. Otherwise, barely a screw remained in place on the old block. Completely new pistons had to be developed. The direct injection system relies on newly-developed jets manufactured by Bosch that had not previously been installed in a production engine.

The Discovery was also offered with the 3.5 litre V8 and later on with the 3.9 litre unit. In 1993, a particularly keenly-priced 2.0 litre petrol version was offered to get round company vehicle tax in the UK which hit vehicles priced over £20,000 very hard. Company cars make up a very high percentage of vehicles in the UK market. The 2.0 litre engine had a respectable sounding 134bhp but its maximum torque figure of 137Nm at 2500rpm was far behind that of the diesel.

In 1993, the Discovery also entered the Japanese market where as part of an agreement with the then Rover shareholder Honda, the Discovery was marketed as the Honda Crossroad. For probably the first time, a Japanese car manufacturer put its emblem on a foreign product for the domestic market. By now the Discovery was a runaway success, not least because it was kept alive by a constant programme of updates which of course hadn't been the case with the Range Rover. In 1990, the four-door version was introduced and was followed in the autumn of 1992 by a V8 automatic using the Range Rover's ZF gearbox. From the 1994 model year onwards this automatic was also available for the diesel. Concurrently the Discovery was facelifted, the interior was given a spring clean and the 3.5 litre V8 replaced by the 3.9 litre version. Research in the USA prior to the Discovery's introduction to the North American market had an influence on the vehicle's badging. At that stage, the name Discovery still appeared on the leading edge of the bonnet. After a series of clinics held in New York and California it was decided that the Land Rover name should appear on the bonnet. In Germany, a driver airbag was now standard equipment and a passenger airbag was an optional extra. ABS had been brought on board the previous year. From 1997 onwards, ABS and twin airbags were also made available. Electric windows and electrically adjustable wing mirrors could be ordered from the factory at no extra cost. The facelift did the Discovery a massive favour. The front end with its enlarged headlights and Land Rover lettering across the bonnet makes it look rather more balanced and elegant than the ungainly snout the vehicle had been given at its launch. In 1995, the diesel automatic version was given more power in the form of an electronically regulated diesel control system manufactured by Bosch which raised the output to 122bhp. This was to cancel out the loss in power brought about by the torque converter. An added bonus of the change was that it increased the maximum torque figure to 300Nm at 2000rpm. This made the automatic Discovery a particularly attractive option. Top speed remained the same at about 97mph.

In contrast to the Defender which, apart from being a functional utility vehicle, has also profited from its increasing perception in the market place as a cult vehicle, the Discovery has had to fight the good fight against some extremely good Japanese opposition. Up until now and particularly in the UK market, the Discovery has been very

New goings on inside: A Driver airbag has been standard since 1994

successful, competing directly with the Mitsubishi Shogun (known in other markets as a Montero or a Pajero) which was similarly priced. The more powerful V8 version requires a bigger bank balance, however.

After eight years in production and only one facelift, it will soon be time for a change. It will come at the end of 1998. The new vehicle will be 15 centimetres longer at the back while retaining the same wheelbase, thus creating more luggage space. There is speculation about the engines being offered. Questions are also being asked about whether the fourth child in the Rover household will lead to any cannibalisation in the range of vehicles. However, the Discovery is a big boy these days and can cope with one or more derivatives. The most important thing is that it stays within the family.

As you like it: A range of options is available to customise the Discovery, which is now in its ninth year of production

Desert storm: Discovery is one shopping trolley that really can off-road

The Future beckons

Bavarians and Freelanders

The whipping boys: One of the last hand-built pre-production test vehicles next to a converted Maestro van with Freelander running gear

Rover boss Walter Hasselkus proclaimed in the Spring of 1997, "the world is waiting for British cars". The new boss of the BMW subsidiary certainly seemed very optimistic. "Subsidiary" is not a word the Brits like to hear, particularly when one considers that Rover's new masters were the very same people who, eight years before the launch of the Series I, had intended flattening London.

Was it all really in the past? Isn't BMW Germany? Obviously, BMW represents Germany (especially in British eyes). As well as this, the past is very much part of the present. Some of the oldest of Rover's 39,000 strong work force experienced the war and particularly the difficult post-war period at first hand. Younger employees have certainly learned all about it from their parents and grandparents. Birmingham and the Midlands were, along with London, subjected to many intensive air raids by the German air force. BMW may well have underestimated the feeling of resentment at the takeover of the company by a former enemy and that despite all the political and cultural progress that has occurred in the years since World War II, Germans still have the burden of their role as aggressor in the conflict on their shoulders and the phrase "we have enough Germans around here" was often heard in Rover management circles after the takeover. BMW still has to tread carefully in Solihull, (if not throughout the Rover group) if the experiment to transform its Rover subsidiary into a company that will build more cars than the parent company itself is to succeed. It was not for nothing that the Munich bosses left an Englishman John Towers in charge of Rover during the immediate period after the takeover. As it increasingly became clear that the differences in Tower's management style could not be

FREELANDER™

overlooked by the Bavarians, a BMW man was brought in to do things the BMW way, albeit in a manner sensitive to British sensibilities. The refined Hasselkus who, as boss of BMW GB and BMW South Africa, has considerable experience of the Anglo-Saxons, was obviously the right man for the job. Before he joined Rover, he had been responsible for putting BMW's Motorbike division back on course.

Despite a sales drop in the domestic market, Rover had produced a total of 507,300 vehicles at three sites; Longbridge (100, 200, 400, Mini and MGF), Cowley (600 and 800) and obviously Solihull. This was the highest ever total recorded by the company and Land Rover's contribution was a record 127,000 vehicles. Both Land Rover and Rover Group as a whole, are the beneficiaries of considerable investment which will lead to significant growth in the future. It is planned that by 2002 230,000 Land Rovers will be coming off the production lines per year.

The key to this dramatic increase in production is Land Rover's new baby, for which a capacity of at least 60,000 vehicles a year is foreseen. Freelander has had a new production facility built for it costing £241m. A major challenge during the initial pre-production stage was guaranteeing being able to deliver the highest possible quality levels. In recent years, there have been many complaints by Discovery owners – the manual gearbox in particular has been a source of much dissatisfaction. One can be certain that BMW, which has built its reputation on quality, will use its influence to remedy the situation. Jaguar had been affected by a similar malaise and has since managed to get its act together. So will Land Rover. Solihull cannot rely on the loyalty of its fans and their willingness to tolerate quality shortcomings to an extent which is unusual in the motor industry, for much longer.

The BMW takeover is changing the way things are done within Rover Group and the Freelander represents a new dawn for Land Rover. First of all, Freelander is not an off-roader in the classic mould. One could

Breaking new ground: Freelander is designed to take the small 4x4 leisure car class by storm. This is the five-door Station Wagon

On the ramp: The new boy has just passed its initiation rites on the test track

assigned solely to computers. Only when the Freelander existed as a three-dimensional virtual image did the engineers start on constructing the first prototype. One of the advantages of this method of design and build is that the vehicle and its tooling can be developed concurrently and therefore more quickly.

The 500 strong development team was led by Richard Elsy. It is an open secret these days that many observers, even within Land Rover, viewed the Freelander team building such a new and revolutionary Land Rover with somewhat sceptical eyes because it really was a venture into the technically unknown. Elsy has gone on record though, saying that even the project's greatest doubters were finally won over when they saw the final product and experienced what it was capable of. No other Land Rover has ever been so competent on the road although the Range Rover with its technically advanced suspension system comes close.

Fleet of foot: The Freelander is particularly agile in the rough

At the same time, the off-road capability of the vehicle was not ignored. The independent suspension is stronger than everything else in its class and more than equal to the forces it is likely to face. Finally, Elsy pointed out that in terms of build quality, the Freelander will also set new standards. This is one of the reasons why the pre-production phase has been relatively long.

With this family of four, Land Rover hardly needs to be worried about the future. In actual fact the opposite is true. There is a lot more to come. It would seem that BMW has laid the foundation for a new golden age. Consequently Munich's plans for its own off-roader do not appear to sit too comfortably with Land Rover's hopes and intentions. Nevertheless, a vast amount of money is being invested in the Solihull plant. At Gaydon, next door to the Motor Heritage Museum just to the south of Birmingham, Rover opened its new research and development facility in the spring of 1997. The test facility gives Rover and Land Rover the most up-to-date development and test technology there is available today. Plans for

Fashion statement: Styling cues for the space age make Land Rover's baby ready to party

the development of this crucial facility were already well in hand when the BMW takeover was announced.

At Hams Hall near Birmingham, Rover is currently building a new engine plant which is costing over £400m. The plant will manufacture engines for both Rover and BMW cars. Recently a brand new £110m paint shop was opened at Solihull. This is currently the largest paint facility in Europe. It must not be forgotten however that BMW wants a return on their investment in Rover. This is unlikely to happen much before the year 2000. Land Rover is the only division that consistently makes a profit. At Rover Group, the total company results still leave the company in the red. In this respect, little has changed over the last thirty years. Over the last fifty years however, the confidence of the Land Rover workforce has grown enormously. Just imagine for a moment if Maurice Wilks had not bothered to look for a substitute for his tired old World War II Jeep. Would Rover have survived until today without Land Rover?
Probably not.

Funtastic: Freelander Softback with special equipment. There are 100 accessories for every possible activity

Designer detail: The Land Rover logo crops up in the third brake light

The Landy is alive

What to know about the Land Rover scene

The Internet is a veritable treasure trove of information for Land Rover fanatics. www.landroverworld.com is a web site set up by Rover Group

If the Land Rover were not a classic cult vehicle there would be no clubs, books, magazines and Internet sites celebrating its existence. Numerous examples of the marque are in museums across the globe and fans can have every kind of accessory from caps to model cars.

In virtually every country on the planet there is a Land Rover club. In the UK alone there are over 180 some of which are dedicated to particular models be they "Series" vehicles, Range Rovers or Discoverys. 37 of these are grouped under the umbrella of the "Association of Rover Clubs Ltd" (ARC). Many more are not part of this organisation. People meet to talk Land Rover, swap tales, hunt for spare parts, plan expeditions both on road and off. In Germany, the off-roading thing is rather more difficult than in the UK as German state bureaucracy has made it virtually impossible to drive off-road anywhere. In the UK however, it is still a different story and there is ample opportunity to take Land Rovers far away from the black stuff. Additionally, there are 135 off road schools where one can get the Land Rover really muddy without running the risk of upsetting anybody. It goes without saying that a visit to Great Britain is an absolute must for any Land Rover fan – and you don't have to go straight to Anglesey. Land Rovers are a part of the local scenery everywhere you go. It is astonishing just how many pre-1971 models are still running around the British Isles with their classic Land Rover face. Getting hold of a cheap, used Land Rover in the UK is easy and the range of real classics on offer is obviously bigger than anywhere else.

The marque even has two dedicated monthly magazines which can be bought in virtually every newsagent in the country (and abroad) Land Rover World and Land Rover Owner International. If you are looking for an unusual Land Rover, the classifieds at the back of both of these magazines are a rich source of possible candidates. The number of after market suppliers for the Land Rover community is also astounding, with everything one could possibly imagine bolted to your Land Rover being offered for sale, from winches for Land Rovers to spray guards for the Discovery and even a third childrens' bench seat for the Range Rover.

One journey every Land Rover aficionado should make is to the Dunsfold Land Rover trust in Surrey. 70 year old Brian Bashall has never driven anything but Land Rovers. Over the years Brian, who is about as passionate about Land Rovers as it is possible to get, has collected over fifty very unusual examples of the marque. Fortunately, he displays them to the general public. Visitors need time to view the collection however as the vehicles are not all at one location and live in a number of garages. He doesn't charge a fee. Among the more unusual variants are a couple of two-wheel drive Series IIIs ordered by the Belgian Army. Brian is particularly proud of his collection of 101 Forward Controls (his son Philip was also responsible

Everything that the enthusiast wanted to ask: In the UK two dedicated magazines satisfy the aficionado's need for information. Rover Blatt is the German Land Rover Club's magazine

for manufacturing the vehicles used in the Judge Dredd film with Sylvester Stallone which were based on ex-army 101s). The museum is financed by the Bashall family itself and a trust to which members contribute £20 per year. Additionally, he sells the odd vehicle from the collection every once in a while but does this with a heavy heart. It was just such a sale that made the day of an American collector who acquired a very pretty 80inch Series I, the first ever Lightweight and an SAS Pink Panther in the spring of 1997.

Another must which concentrates not just on Land Rovers, but on the British Motor Industry as a whole is the Heritage Motor Centre in Gaydon just south of Birmingham. This mecca of the motor industry in the UK encompasses the complete history of the British motor car. On the land surrounding the museum, there is also an extensive off-road course which enables Land Rovers to demonstrate what they can do away from the tarmac.

For the non-automotively-minded, Land Rovers are associated with the TV series Daktari (if they are capable of distinguishing a Land Rover from a Jeep at all). Fans of the two classic off-roaders are actually very different from one another. Land Rover fans tend to be fairly laid back and rarely push their vehicles to the limit. Some Land Rover clubs in the UK are only for individual marques. The Series I Club produces a high quality newsletter which is dedicated to the preservation of the grandad of all Land Rovers. The Lightweight Club is of a similar ilk, and appeals to the fans of the military half tonne which has acquired something of a cult status among Land Rover owners. Other clubs such as the Cornwall and Devon Land Rover Club, have established a formidable reputation within the club scene for their driving abilities in off-road trialling competitions. Outside of the ARC, the club scene is just as lively with a slightly less formal approach to club life. Abroad, Land Rover clubs are often more oriented to the off-road expeditioning aspect of Land Rover ownership. A club like the Land Rover Owners' Club of Southern Africa organises numerous expeditions into the bush and regularly holds lectures on how to prepare vehicles for the task. An example of a club which manages to combine both aspects is the Deutscher Rover Club (DRC). In existence since 1975, the club covers everything possible to do with Land Rovers. Despite the name "Rover Club" it must be emphasised that this is indeed a Land Rover club as Rover cars hardly feature at all. When the club was founded, the Germans did not make the distinction between Rovers and Land Rovers. Initially, the club was set up by

These are the winners (well almost): Participants in the international selections in the United Arab Emirates in 1997

a group of Range Rover owners who were unhappy with the provision of spare parts for their vehicles and felt they could do better themselves. Today, the club numbers over 800 and members drive everything from Defenders to Discoverys, from Series Is to Lightweights and Forward Controls to Range Rovers. Many members are currently giving up their Series IIIs to buy Defenders according to club spokesperson Roland Koch. The DRC sees itself on the one hand as a classic car club, organising spare part runs to the UK, and also as a forum for those considering long-distance expeditions and off-roading events, all of which are discussed at regular club meetings.This club is divided into eleven regional groupings much like Germany itself is divided into federal states. Interestingly, there aren't many members from the former German Democratic Republic. Before 1989 it was virtually impossible to drive a Land Rover behind the wall, unless your name was Erich Honecker that is. The former leader of the socialist East German state had a number of Land Rovers specially converted for hunting expeditions.

The club magazine Rover Blatt is a worthy production appearing six times a year. Over half of the club's budget goes into producing the magazine. Twice a year, it organises national meetings and in 1998 the DRC wants to organise a large celebration near the Nuerburgring race circuit as part of the fiftieth anniversary celebrations, at which over 400 vehicles are expected. The club also intends to be present at the Land Rover 50th Anniversary celebrations in April and May 1998. In 1998, half the world will be celebrating Land Rover's birthday. The British Series I club is planning a large meeting on the last weekend in June at Shugborough Hall near Stafford. In Australia, the Victoria Land Rover Owners' Club is to undertake a fourteen day drive through the challenging Simpson desert to show that the older Land Rovers are more than capable of coping with everything thrown at them. Participants are only allowed to take part in Series I and II vehicles. In Austria, Land Rover fans are spread across a number of locally-organised clubs. In Switzerland, aficionados have been meeting for ten years as the Land Rover Club Switzerland. The 180 members celebrated

In 1985 the teams had to struggle through the jungles of Borneo

their tenth anniversary in the Bernese Oberland in the summer of 1997. The Swiss are known for their cosmopolitan approach to matters and consequently invited Land Rover owners from all over Europe.

The Camel Trophy event is also intrinsically linked to Land Rover. It is however, not just another operation by Land Rover's marketing department. It was actually an event that came about through the efforts of the tobacco firm Reynolds' German subsidiary to strengthen the image of Camel cigarettes. The first event was staged in 1980 with three German teams crossing through the Amazon using Jeeps. Today the event is a professionally-organised affair based on the promotion of Camel Trophy adventure wear and the Camel cigarette part of the equation has managed to move into the background. From 1982 on, the event went international and from then on anyone who felt they were fit enough to complete the event could apply. National and international selection camps decide the ultimate teams for the Trophy event itself. Every year the number of applicants runs into the thousands. In 1997, twenty teams took part in the event in Mongolia. The teams took part driving Discoverys and were supported by two Defenders. In 1998, the Trophy is to take place in Tierra del Fuego in South America, this time using Freelanders. Land Rover has supported the event since 1981 and Camel has to pay for its vehicles. Range Rovers, Defenders and Discoverys have all demonstrated their abilities on this gruelling event. 17 years of involvement have led to the Camel Trophy being automatically associated with Land Rover. The association doesn't work the other way however. Very few people think of Camel Trophy when they are given the name Land Rover.

Whether Camel Trophy fans, or dedicated to restoring their prize Series I, Land Rover owners are notoriously difficult to be lumped into one all-encompassing category and many of them are more than a little bit wary of any form of club, even when it is about their beloved Land Rover. Land Rover owners are individualists and this is a quality which doesn't exactly fit well with club life. At the same time, the loners among the Land Rover fans do actually feel drawn to other Land Rover owners because there is a common bond. This seems to be stronger in the case of the Land Rover than for other makes of vehicle. It is difficult to explain. Perhaps it has something to do with the aura that surrounds the Land Rover. The Land Rover has an original feel to it, smacking of adventure and making its history immediately obvious to anyone who sees one. It is a child of the post-war period, like many of its drivers, it isn't perfect but like many people it tries its best. And that is why we like it so much.

River crossings such as this in Australia in 1986 are one of the easier tasks that Land Rovers are presented with

For the 1998 Camel Trophy it is the Freelander's turn to receive its call-up papers

Hard to swallow

Technical specifications* and production figures

*Land Rover technical data often includes contradictory information. On the following pages the authors have used primary sources where necessary. Information refers to the most important models

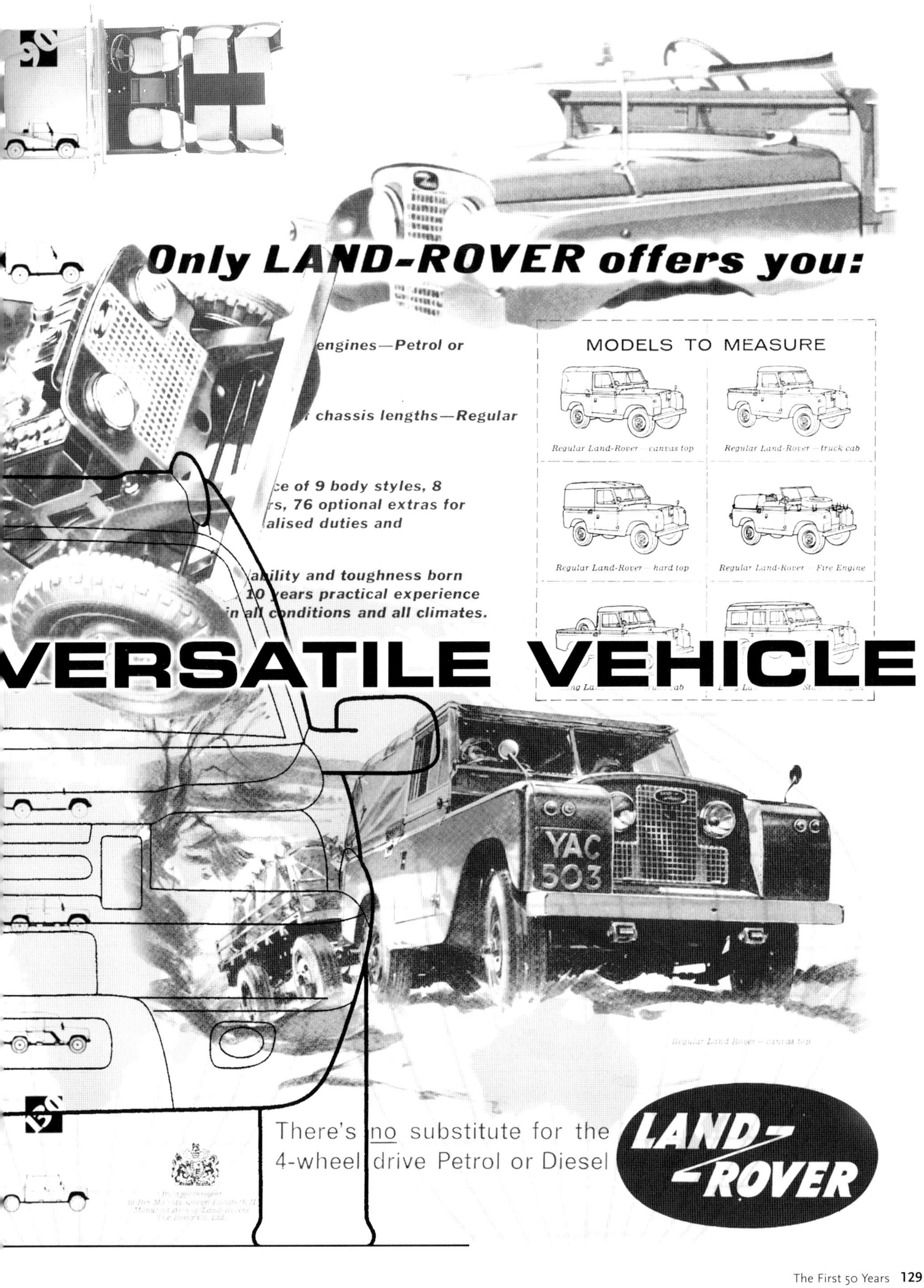
Only LAND-ROVER offers you:
engines—Petrol or
chassis lengths—Regular
ce of 9 body styles, 8
rs, 76 optional extras for
alised duties and
ability and toughness born
10 years practical experience
in all conditions and all climates.
MODELS TO MEASURE
Regular Land-Rover — canvas top
Regular Land-Rover — truck cab
Regular Land-Rover — hard top
Regular Land-Rover — Fire Engine
VERSATILE VEHICLE
YAC 503
There's no substitute for the
4-wheel drive Petrol or Diesel
LAND-ROVER

Series I
1948 until 1958

Two-door Pick-Up with canvas tonneau or two-door Station Wagon (Oct. 1948 until Spring 1951). 1.6 litre, four cylinder engine, IOE (overhead inlet valves/side exhaust valves);
1950: Hardtop for Pick-Up;
1952: 2.0 litre engine;
1954: four-door Station Wagon, Pick-Up with longer wheelbase (107 inch);
all short-wheelbase-models: 86 inch wheelbase; new instruments

	80 1.6 litre engine 1948 until 1952	**86** 2.0 litre engine from 1954	**107 Station Wagon** 2.0 litre engine from 1955
Engine:	Water-cooled four cylinder engine, Solex carburettor	Water-cooled four cylinder engine, Solex carburettor	Water-cooled four cylinder engine, Solex carburettor
Cylinder Capacity:	1595cc	1997cc	1997cc
Bore x Stroke:	69.5mm x 105mm	77.8mm x 105mm	77.8mm x 105mm
Compression:	6.8:1	6.8:1	6.8:1
Power output:	50bhp/4000rpm	52bhp/4000rpm	52bhp/4000rpm
Torque:	108Nm/2000rpm	137Nm/1500rpm	137Nm/1500rpm
Power train:	Four-speed gearbox with reverse and transfer box; third and top gear synchromeshed; permanent four-wheel-drive with freewheel in front drive. From 1950: Selectable four-wheel- or two-wheel-drive (rear) in high range; permanent four-wheel-drive in low range (no freewheel)	Four-speed gearbox with reverse and transfer box; third and top gear synchromeshed; selectable four-wheel- or two-wheel-drive (rear) in high range; permanent four-wheel-drive in low range (no freewheel)	Four-speed gearbox with reverse and transfer box; third and top gear synchromeshed; selectable four-wheel- or two-wheel-drive (rear) in high range; permanent four-wheel-drive in low range (no freewheel)
Gears:	3.00:1, 2.04:1, 1.47:1, (from 1950: 1.38:1), 1:1, R 2.54:1; transfer box giving 1.148:1 or 2.89:1 step-down ratio in high/low range	3.00:1, 2.04:1, 1.38:1, 1:1, R 2.54:1; transfer box giving 1,148:1 or 2,89:1 step-down ratio in high/low range	3.00:1, 2.04:1, 1.38:1, 1:1, R 2.54:1; transfer box giving 1,148:1 or 2,89:1 step-down ratio in high/low range
Wheels:	6.00 x 16, 7.00 x 16 or 6.50 x 16	6.00 x 16 or 7.00 x 16 or 6.50 x 16	6.00 x 16 or 7.00 x 16 or 7.50 x 16
Suspension:	Live front and rear axle with semi-elliptic leaf springs and hydraulic telescopic shock absorbers	Live front and rear axle with semi-elliptic leaf springs and hydraulic telescopic shock absorbers	Live front and rear axle with semi-elliptic leaf springs and hydraulic telescopic shock absorbers
Brakes:	Hydraulic brakes (drums)	Hydraulic brakes (drums)	Hydraulic brakes (drums)
Steering:	Worm and nut with recirculating ball	Worm and nut with recirculating ball	Worm and nut with recirculating ball
Turning circle:	10.7 or 12.2m	11.27 or 12.2m	15.2m
General technical specification:			
Wheelbase:	2.03m	2.18m	2.72m
Track:	1.27m	1.27m	1.27m
Length:	3.35m	3.57m	4.41m
Width:	1.55m	1.59m	1.59m
Height:	1.79m (with canvas top)	1.94m (with canvas top)	1.98m
Topspeed:	ca. 55mph	ca. 55mph	ca. 55mph
Unladen weight:	1174kg	1224kg	1560kg
Max. payload:	453kg	453kg	680kg

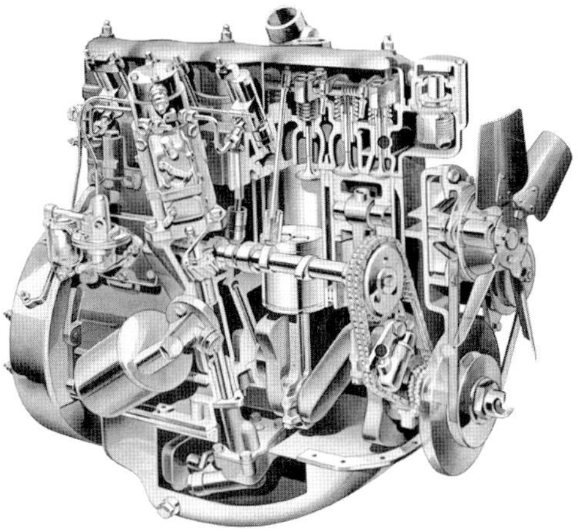

Cutaway diagram of the first Land Rover diesel engine, offered since 1957

Rear power take-off pulley

88 2.0 litre diesel engine from 1957	109 2.0 litre engine from 1957
Water-cooled four cylinder diesel engine	Water-cooled four cylinder engine, Solex carburettor
2052cc	1997cc
85.7mm x 88.9mm	77.8mm x 105mm
22.5:1	6.8:1
51bhp/3500rpm	52bhp/4000rpm
118Nm/2000rpm	137Nm/1500rpm
Four-speed gearbox with reverse and transfer box; third and top gear synchromeshed; selectable four-wheel- or two-wheel-drive (rear) in high range; permanent four-wheel-drive in low range (no freewheel)	Four-speed gearbox with reverse and transfer box; third and top gear synchromeshed; selectable four-wheel- or two-wheel-drive (rear) in high range; permanent four-wheel-drive in low range (no freewheel)
3.00:1, 2.04:1, 1.38:1, 1:1, R 2.54:1; transfer box giving 1,148:1 or 2,89:1 step-down ratio in high/low range	3.00:1, 2.04:1, 1.38:1, 1:1, R 2.54:1; transfer box giving 1,148:1 or 2,89:1 step-down ratio in high/low range
6.00 x 16 or 7.00 x 16	7.00 x 16 oder 7.50 x 16
Live front and rear axle with semi-elliptic leaf springs and hydraulic telescopic shock absorbers	Live front and rear axle with semi-elliptic leaf springs and hydraulic telescopic shock absorbers
Hydraulic brakes (drums)	Hydraulic brakes (drums)
Worm and nut with recirculating ball	Worm and nut with recirculating ball
11.6m	13.7m
2.24m	2.77m
1.27m	1.27m
3.57m	4.41m
1.59m	1.59m
1.92m (with canvas top)	2.11m (with canvas top)
ca. 55mph	ca. 60mph
1330kg	1395kg
611kg	680kg or 906kg (with bigger wheels)

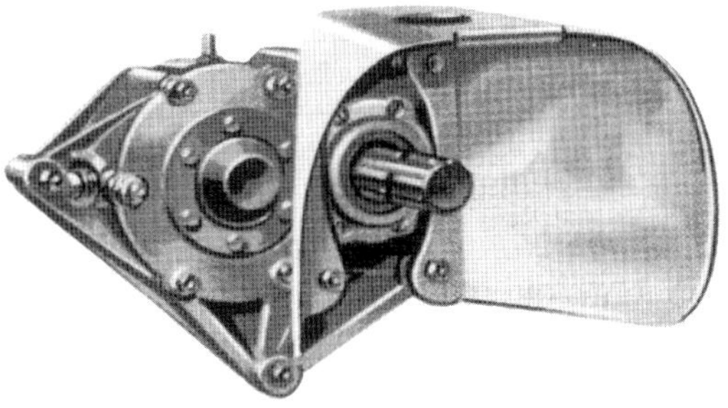

Rear power take-off drive

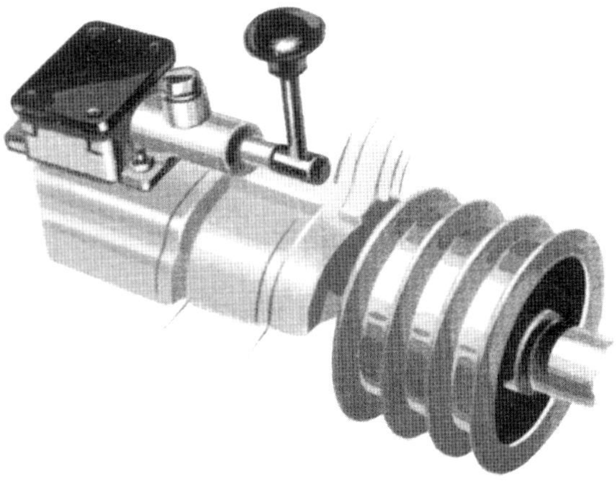

Center power take-off with control lever

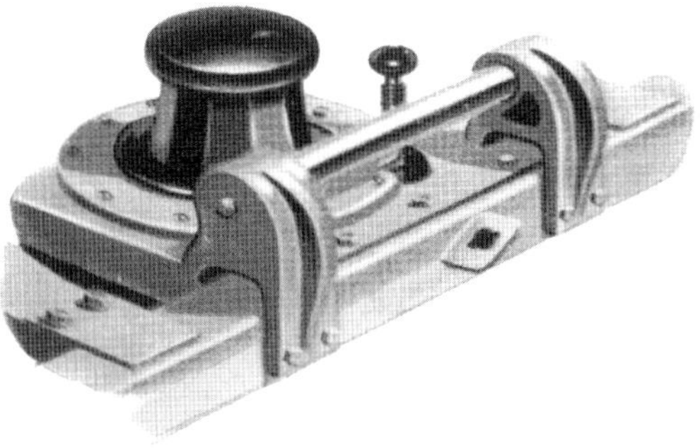

The capstan winch mounted behind the front bumper was driven by the crankshaft and could pull up to 1150kg

Series II
1958 until 1961

Built only in a three-year-production run, the Series II featured changes in styling. Cylinder capacity of the petrol engine increased. Two-door Pick-Up with canvas tonneau or two- resp. four-door Station Wagon. Two engine options to choose.

Series II A
1961 until 1971

Limited changes made to the Series II included the following: A diesel engine with the same capacity as the petrol engine. 1967 saw the introduction of a six cylinder petrol engine and a new gearbox. In 1968, the 109inch 1 ton variant was the first Land Rover capable of carrying more than 1000kg. The headlights were moved from their position in the grille to a new location in the wings.

	88 2.3 litre engine from 1958	**109** 2.0 litre diesel engine from 1958	**88** 2.25 litre diesel engine from 1961
Engine:	Water-cooled four cylinder engine, Solex carburettor	Water-cooled four cylinder diesel engine	Water-cooled four cylinder diesel engine
Cylinder Capacity:	2286cc	2052cc	2286cc
Bore x Stroke:	90.5mm x 88.9mm	85.7mm x 88.9mm	90.5mm x 88.9mm
Compression:	7.0:1	22.5:1	23:1
Power output:	77bhp/4250rpm	51bhp/3500rpm	62 bhp/4000rpm
Torque:	168/2500rpm	118Nm/2000rpm	140Nm/1800rpm
Power train:	Four-speed gearbox with reverse and transfer box; third and top gear synchromeshed; selectable four-wheel- or two-wheel-drive (rear) in high range; permanent four-wheel-drive in low range (no freewheel)	Four-speed gearbox with reverse and transfer box; third and top gear synchromeshed; selectable four-wheel- or two-wheel-drive (rear) in high range; permanent four-wheel-drive in low range (no freewheel)	Four-speed gearbox with reverse and transfer box; third and top gear synchromeshed; selectable four-wheel or two-wheel-drive (rear) in high range; permanent four-wheel drive in low range
Gears:	3.00:1, 2.04:1, 1.38:1, 1:1, R 2.54:1; transfer box giving 1,148:1 or 2,89:1 step-down ratio in high/low range	3.00:1, 2.04:1, 1.38:1, 1:1, R 2.54:1; transfer box giving 1,148:1 or 2,89:1 step-down ratio in high/low range	3.00:1, 2.04:1, 1.38:1, 1:1, R 2.54:1; transfer box giving 1.148:1 (high) or 2.89:1 (low)
Wheels:	7.50 x 16	7.50 x 16	7.50 x 16
Suspension:	Live front and rear axle with semi-elliptic leaf springs and hydraulic telescopic shock absorbers	Live front and rear axle with semi-elliptic leaf springs and hydraulic telescopic shock absorbers	Live front and rear axles with semi-elliptic leaf springs and hydraulic telescopic shock absorbers
Brakes:	Hydraulic brakes (drums)	Hydraulic brakes (drums)	Hydraulic brakes (drums)
Steering:	Worm and nut with recirculating ball	Worm and nut with recirculating ball	Worm and nut with recirculating ball
Turning circle:	11.6m	13.7m	11.6m
General technical specification:			
Wheelbase:	2.24m	2.77m	2.24m
Track:	1.31m	1.31m	1.31m
Length:	3.62m	4.45m	3.62m
Width:	1.63m	1.63m	1.63m
Height:	1.97m (with canvas top)	2.05m (with canvas top)	1.97m (with canvas top)
Topspeed:	ca. 70mph	ca. 60mph	ca. 60mph
Unladen weight:	1314kg	1580kg	1402kg
Max. payload:	703kg	1094kg	615kg

Series III 1971 until 1985

Slight changes to technical specifications and configurations. The most noticeable change is the plastic radiator grille. New was an all-synchromesh gearbox, larger brakes, clutch, a new dashboard. The first gear ratio was now lower (3.68, and later on 3.73:1) In 1979, the V8 engine from the Range Rover finally replaced the six cylinder. The V8 variants were also equipped with permanent four-wheel-drive.

109 2.6 litre engine from 1967	**109 1-ton Pick-Up** 2.6 litre engine from 1968	**109 V8 Station Wagon** 3.5 litre engine from 1979
Water-cooled six cylinder engine, Zenith carburettor 2625cc 77.8mm x 92.1mm 7.8:1 83bhp/4500rpm 172Nm/1500rpm	Water-cooled six cylinder engine, Zenith carburettor 2625cc 77.8mm x 92.1mm 7.8:1 83bhp/4500rpm 173Nm/1500rpm	Water-cooled V8 cylinder engine, two Zenith-Stromberg carburettors 3528cc 88.9mm x 71.1mm 8.1:1 91bhp/3500rpm 225Nm/2000rpm
Four-speed gearbox with reverse and transfer box, third and top gear synchromeshed; selectable four-wheel- or two-wheel-drive (rear) in high range; permanent four-wheel-drive in low range 3.60:1, 2.22:1, 1.50:1, 1:1, R 3.02:1; transfer box giving 1.148:1 (high) or now 2.35:1 (low) 7.50 x 16	Four-speed gearbox with reverse and transfer gearbox; third and top gear synchromeshed; selectable four-wheel- or two-wheel-drive (rear) in high range; permanent four-wheel-drive in low range 3.60:1, 2.22:1, 1.50:1, 1:1, R 3.02:1; transfer box giving 1.53:1 (high) or 3.27:1 (low) 9.00 x 16	Four-speed gearbox with reverse and transfer box; permanent four-wheel-drive with a lockable centre differential 4.07:1, 2.45:1, 1.51:1, 1:1, R 3.66:1; transfer box giving 1.33:1 (high) and 3.32:1 (low) 7.50 x 16
Live front and rear axle with semi-elliptic leaf springs and hydraulic telescopic shock absorbers Hydraulic brakes (drums) Worm and nut with recirculating ball 13.7m	Live front and rear axle with semi-elliptic leaf springs and hydraulic telescopic shock absorbers Hydraulic brakes (drums) Worm and nut with recirculating ball 13.7m	Live front and rear axle with semi-elliptic leaf springs and hydraulic shock absorbers Hydraulic brakes (drums) Worm and nut with recirculating ball 13.7m
2.77m 1.31m 4.45m 1.63m 2.05m (with canvas top) ca 75mph 1566kg 1108kg	2.77m 1.31m 4.45m 1.63m 2.05m (with canvas top) ca. 60mph 1760kg 1297kg	2.77m 1.33m 4.45m 1.68m 2.01m 80mph 1682kg 1028kg

90/110/Defender
1983 until present

As before, a comprehensive model range (long wheelbase, short wheelbase, open, closed, Pick-Up). The radiator is moved forward. A one-piece windscreen is introduced, coil-springs instead of leaf springs, permanent four-wheel-drive. As time progresses power-assisted steering, disc brakes are introduced to the range. In 1985, the older four cylinder petrol engines were finally dropped from the model range (although Land Rover still builds the engines). The base engine is now the direct injection 2.5 litre diesel. From 1990, the model name Defender is introduced as part of moves to tighten the corporate identity. The wheelbase was increased to 110 inches, the short wheelbase is named the 90, although its actual wheelbase is closer to 93 inches. The Crew-Cab is a new variant – a Pick-Up with a double cab and an extended wheelbase of 127 inches (subsequently named the 130). The V8 engine is still available in some markets. The Defender is capable of towing 3.5 tons (with air-brake assistance, this increases to six tons).

	110 Softtop 2.25 litre engine 1983 until 1985	**90 Station Wagon** 2.5 litre engine 1984 until 1994	**Defender 90 Tdi Softtop** 2.5 litre turbo-diesel engine 1997 Model Year
Engine:	Water-cooled four cylinder engine, Weber carburettor	Water-cooled four cylinder engine, Weber carburettor	Water-cooled four cylinder turbo-diesel engine with direct injection
Cylinder Capacity:	2286cc	2494cc	2495cc
Bore x Stroke:	90.5mm x 88.9mm	90.5mm x 97.0mm	90.5mm x 97.0mm
Compression:	8:1	8:1	19.51:1
Power output:	74bhp/4000rpm	83bhp/4000rpm	113bhp (83kw)/4000rpm
Torque:	163Nm/2000rpm	180Nm/2000rpm	265Nm/1800rpm
Power train:	Four-speed gearbox with reverse and transfer box; permanent four-wheel-drive with a lockable centre differential	Four-speed gearbox with reverse and transfer box; permanent four-wheel-drive with a lockable centre differential	Five-speed gearbox with reverse and transfer box; permanent four-wheel-drive with a lockable centre differential
Gears:	4.07:1, 2.45:1, 1.51:1, 1:1, R 3.66:1; five-speed gearbox (option) 3.59:1, 2.3:1, 1.51:1, 1:1, 0.83:1, R 3.66:1; transfer box giving 1.671:1 (high) or 3.32:1 (low)	4.07:1, 2.45:1, 1.51:1, 1:1, R 3.66:1; optional five-speed gearbox: 3.59:1, 2.3:1, 1.51:1, 1:1, 0.83:1, R 3.66:1; transfer box giving 1.67:1 (high) and 3.32:1 (low)	3.69:1, 2.13:1, 1.40:1, 1:1, 0.77:1; R 3.43:1; transfer box giving 1.41:1 (high) or 3.32:1 (low)
Wheels:	7.50 x 16	6.00 x 16	6.00 x 16
Suspension:	Live front and rear axle with coil springs; telescopic dampers	Live front and rear axle with coil springs, telescopic shock absorbers	Live front and rear axle with coil springs, telescopic shock absorbers
Brakes:	Hydraulic servo-assisted brakes (discs at the front, drums at the rear)	Hydraulic brakes (discs at the front, drums at the rear)	Hydraulic brakes (discs all round)
Steering:	Recirculating ball steering with optional power assistance	Recirculating ball steering with optional power assistance	Recirculating ball steering with power assistance
Turning circle:	12.8m	12.8m	11.7m
General technical specification:			
Wheelbase:	2.79m	2.36m	2.36m
Track:	1.49m	1.49m	1.49m
Length:	4.45m (Station Wagon: 4.58m, Hi-Cap Pick-Up: 4.67m)	3.88m (3.72m with softtop)	3.87m
Width:	1.79m	1.79m	1.79m
Height:	2.03m	1.96m (with softtop)	1.96m (with softtop)
Topspeed:	ca. 70mph	75mph	85mph
Unladen weight:	1591kg	1605kg	1695kg
Max. payload:	1464kg	734kg	705kg

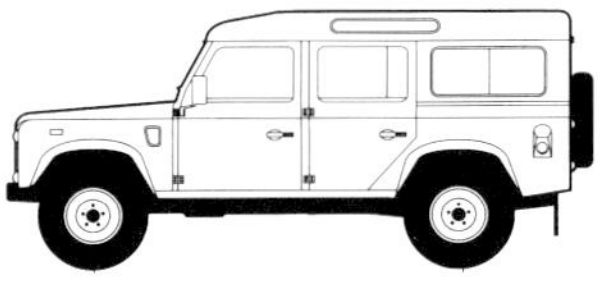

Resisting the wind tunnel. For the fifty years, the basic shape of the Land Rover Defender has not changed

Defender 110 Tdi Station Wagon 2.5 litre turbo-diesel engine 1997 Model Year	**Defender 130 Tdi Crew Cab** 2.5 litre turbo-diesel engine 1997 Model Year
Water-cooled four cylinder turbo-diesel engine with direct injection 2495cc 90.5mm x 97.0mm 19.5:1 113bhp (83kw)/4000rpm 265Nm/1800rpm	Water-cooled four cylinder turbo-diesel engine with direct injection 2495cc 90.5mm x 97.0mm 19.5:1 113bhp (83kw)/4000rpm 265Nm/1800rpm
Five-speed gearbox with reverse and transfer box; permanent four-wheel-drive with a lockable centre differential 3.69:1, 2.13:1, 1.40:1, 1:1, 0.77:1, R 3.43:1; transfer box giving 1.41:1 (high) or 3.32:1 (low)	Five-speed gearbox with reverse and transfer box; permanent four-wheel-drive with a lockable centre differential 3.69:1, 2.13:1, 1.40:1, 1:1, 0.77:1, R 3.43:1; transfer box giving 1.41:1 (high) or 3.32:1 (low)
7.50 x 16	7.50 x 16
Live front and rear axle with coil springs, telescopic dampers Hydraulic brakes (discs all round) Recirculating ball steering with power assistance 12.8m	Live front and rear axle with coil springs, telescopic dampers Hydraulic brakes (discs all round) Recirculating ball steering with power assistance 15.1m
2.79m 1.51m 4.59m	3.23m 1.51m 5.13m
1.79m 2.06m 85mph 2055kg 895kg	1.79m 2.06m 85mph 2086kg 1414kg

Modest in appearances but a real powerhouse. The turbo-diesel engine with direct injection

Four folding individual seats, lap belts. A Defender 90 from the inside

Range Rover 1970 until present

Luxury off-road vehicle "made by Land Rover". From 1981 it was offered with four doors, from 1986, a diesel engine (initially a VM engine, but later Land Rover's own Tdi). This was the first production vehicle with permanent four-wheel-drive. From 1992, it was offered with air suspension. 1994 saw the introduction of the new shape Range Rover with a new chassis and new engines including a BMW diesel.

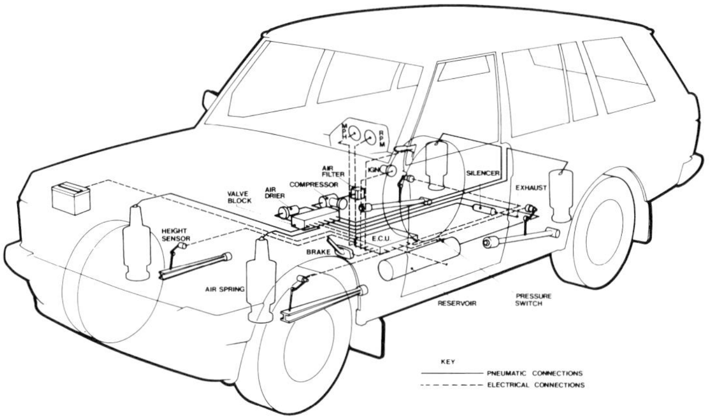

	Range Rover	**Range Rover Turbodiesel (VM)**	**Range Rover 4.2**
	1970 until 1979	1986 until 1992	1994 until 1996
Engine:	Water-cooled V8 cylinder, two Zenith-Stromberg carburettors	Water-cooled four cylinder turbo-diesel engine manufactured by VM in Italy	Water-cooled V8 cylinder engine with Lucas fuel injection
Cylinder Capacity:	3528cc	2393cc	4278cc
Bore x Stroke:	88.9mm x 71.1mm	92.0mm x 90.0mm	94.0mm x 77.0mm
Compression:	8.5:1 (8.3:1 1971-1977, 8.1:1 1977-1979)	22:1	8.9:1
Power output:	135bhp/4750rpm	112bhp/4200rpm	202bhp(149kw)/4850rpm
Torque:	251Nm/2500rpm	248Nm/2400rpm	340Nm/3250rpm
Power train:	Four-speed gearbox with reverse and transfer box; permanent four-wheel-drive, lockable centre differential (1970 with a limited slip differential)	Five-speed gearbox with reverse and transfer box; permanent four-wheel-drive with a lockable centre differential	Four-speed automatic with hydraulic torque converter reverse and transfer box; permanent four-wheel-drive with an automatic lockable centre differential with a viscous-coupling
Gears:	4.07:1, 2.45:1, 1.51:1, 1:1, R 3.66:1; transfer box giving 1.17:1 (high) or 3.32:1 (low)	3.32:1, 2.13:1, 1.39:1, 1:1, 0.77:1, R 3.66:1; transfer box giving 1.19:1 (high) 3.32:1 (low)	2.47:1, 1.47:1, 1:1, 0.72:1, R 2.08:1; transfer box giving 1.2:1 (high) 3.24:1 (low)
Wheels:	205 x 16	205 x 16	7.50 x 16
Suspension:	Live front axle, coil springs, telescopic dampers, radius arms, Panhard rod; live rear axle, coil springs, telescopic dampers, radius arms, A-bracket, self-levelling strut	Live front axle, coil springs, telescopic dampers, radius arms, Panhard rod; live rear axle, coil springs, telescopic dampers, radius arms, A-bracket, self-levelling strut	Live front axle, coil springs, telescopic dampers, radius arms, Panhard rod, air suspension; live rear axle, coil springs, telescopic dampers, radius arms, A-bracket, air suspension, self-levelling strut
Brakes:	Hydraulic brakes (discs all round)	Hydraulic brakes (discs all round)	Servo-assisted brakes (discs all round), ABS
Steering:	Recirculating ball steering; power assistance was offered as an option from 1973	Recirculating ball steering with power assistance	Recirculating ball steering with power assistance
Turning circle:	11.2m	11.2m	11.2m
General technical specification:			
Wheelbase:	2.54m	2.54m	2.74m
Track:	1.49m	1.49m	1.49m
Length:	4.47m	4.47m	4.65m
Width:	1.78m	1.78m	1.82m
Height:	1.78m	1.78m	1.80m
Topspeed:	91mph	85mph	125mph
Unladen weight:	1723kg	1788kg (four door version)	2055kg
Max. payload:	680kg	650kg	465kg

Performance/torque characteristics

Range Rover 4.0

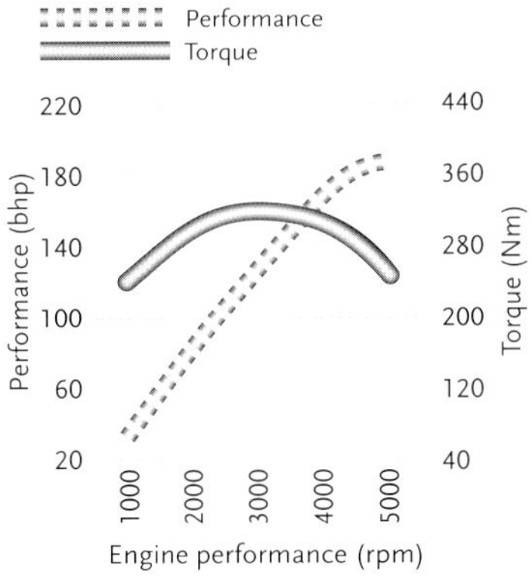

Range Rover 4.6

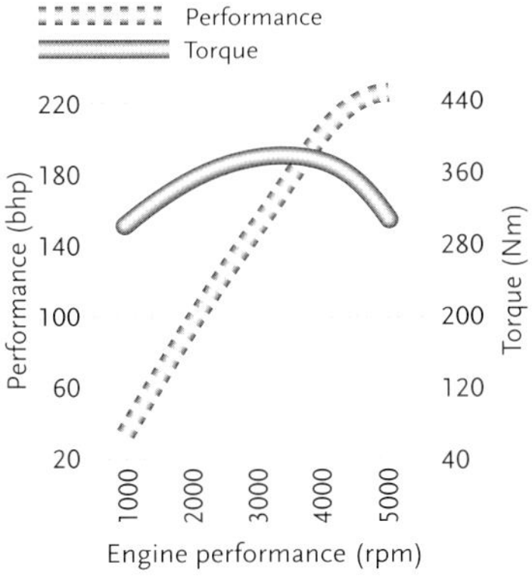

Range Rover 2.5 Diesel

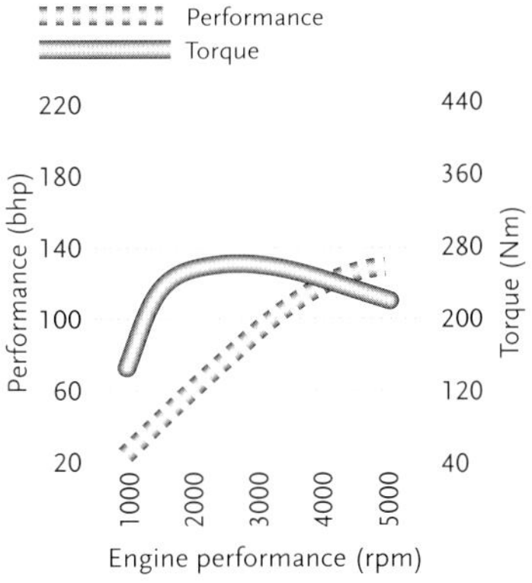

Range Rover 4.0 SE

1994 until present

Water-cooled V8 cylinder engine with electronic fuel injection
3950cc
94.0mm x 71.0mm
89.3:1
186bhp(137kw)/4750rpm
320Nm/3000rpm

Five-speed gearbox (four-speed automatic); permanent four-wheel-drive, automatic locking centre differential with a viscous-coupling

3.32:1, 2.13:1, 1.40:1, 1:1, 0.73:1, R 3.54:1; automatic: 2.48:1, 1.48:1, 1:1, 0.73:1, R 2.09:1; transfer box giving 1.22:1 (high) 3.27:1 (low)
7.00 x 16

Live front axle, coil springs, telescopic dampers, radius arms, Panhard rod, air suspension, anti-roll bar;
live rear axle, coil springs, telescopic dampers, radius arms, air suspension, Panhard rod
Servo-assisted brakes
(discs all round), ABS
Recirculating ball steering
with power assistance

11.85m

2.75m
1.54m
4.71m
1.85m
1.75m to 1.88m
118mph
2090kg
690kg

Range Rover 4.6 HSE

1994 until present

Water-cooled V8 cylinder engine with electronic fuel injection
4533cc
94.0mm x 82.0mm
9.3:1
224bhp(165kw)/4750rpm
380Nm/3600rpm

Four-speed automatic with hydraulic torque converter (ZF) reverse and transfer box; permanent four-wheel-drive with an automatic lockable centre differential with a viscous-coupling
2.48:1, 1.48:1, 1:1, 0.73:1, R 2.09:1; transfer box giving 1.22:1 (high)
3.27:1 (low)

7.00 x 16

Live front axle, coil springs, telescopic dampers, radius arms, Panhard rod, air suspension;
live rear axle, coil springs, telescopic dampers, radius arms, air suspension, Panhard rod
Servo-assisted brakes
(discs all round), ABS
Recirculating ball steering
with power assistance

11.85m

2.75m
1.54/1.53m
4.71m
1.85m
1.75/1.88m
125mph
2220kg
560kg

Range Rover 2.5 DT/DSE

1994 until present

Water-cooled six cylinder turbo-diesel engine (BMW)
2497cc
80.0mm x 82.8mm
22:1
136bhp(100kw)/4400rpm
270Nm/2300rpm

Five-speed gearbox;
permanent four-wheel-drive with an automatic lockable centre differential with a viscous-coupling

3.32:1, 2.13:1,1.4:1 1:1, 0.73:1, R 3.54:1;
transfer box giving 1.22:1 (high)
3.27:1 (low)

7.00 x 16

Live front axle, coil springs, telescopic dampers, radius arms, Panhard rod, air suspension;
live rear axle, coil springs, telescopic dampers, radius arms, air suspension, Panhard rod
Servo-assisted brakes
(discs all round), ABS
Recirculating ball steering
with power assistance

11.85m

2.75m
1.54/1.53m
4.71m
1.85m
1.75/1.88m
105mph
2220kg
560kg

Discovery
1989 until present

Closes the gap between the Land and Range Rover. From 1990, a four-door version was available. In 1992, the 3.9 litre V8 engine was mated to an automatic gearbox, and in 1994 the Tdi was similarly equipped. 1994 also saw a facelift. Towing capacity was 3.5 tons and with air-brakes six tons.

Approach and departure angles

	Discovery V8 1997 Model Year	Discovery 2.5 Tdi 1997 Model Year
Engine:	Water-cooled V8 cylinder engine with Lucas fuel injection	Water-cooled four cylinder turbo-diesel engine with direct injection
Cylinder Capacity:	3947cc	2495cc
Bore x Stroke:	94.0mm x 71.1mm	90.5mm x 97.0mm
Compression:	9.4:1	19.5:1
Power output:	182bhp(134kw)/4750rpm	113bhp(83kw)/4000rpm (manual) or 122bhp(90kw)/4000rpm (automatic)
Torque:	312Nm/3100rpm	265Nm/1800rpm (manual) or 300Nm/2000rpm (automatic)
Power train:	Five-speed gearbox (four-speed automatic); permanent four-wheel-drive with a lockable centre differential	Five-speed gearbox (four-speed automatic); permanent four-wheel-drive with a lockable centre differential
Gears:	3.32:1, 2.13:1, 1.4:1, 1:1, 0.73:1, R 3.43:1; automatic: 2.48:1, 1.48:1, 1:1, 0.73:1, R 2.09:1; transfer box giving 1.22:1 (high) 3.32:1 (low)	3.69:1, 2.13:1, 1.40:1, 1:1, 0.77:1, R 3.43:1; automatic: 2.48:1, 1.48:1, 1:1, 0.73:1, R 2.09:1; transfer box giving 1.22:1 (high) 3.32:1 (low)
Wheels:	235 x 16	205 x 16
Suspension:	Live front axle, coil springs, telescopic dampers, radius arms, Panhard rod, anti-roll bar; live rear axle, coil springs, telescopic dampers, radius arms, A-bracket, anti-roll bar	Live front axle, coil springs, telescopic dampers, radius arms, Panhard rod, anti-roll bar; live rear axle, coil springs, telescopic dampers, radius arms, A-bracket, anti-roll bar
Brakes:	Servo-assisted brakes (discs all round)	Servo-assisted brakes (discs all round)
Steering:	Recirculating ball steering with power assistance	Recirculating ball steering with power assistance
Turning circle:	11.9m	11.9m
General technical specification:		
Wheelbase:	2.54m	2.54m
Track:	1.49m	1.49m
Length:	4.52m	4.52m
Width:	1.79m	1.79m
Height:	1.92m	1.92m
Topspeed:	115mph	97mph
Unladen weight:	1925kg	1935kg
Max. payload:	795kg	785kg

Performance/torque characteristics Discovery 2.5 Tdi

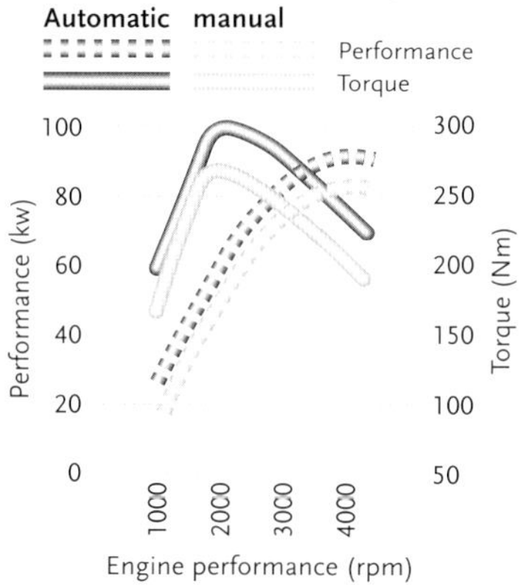

Thanks to an improved injection system, the Tdi automatic is superior to the manual version both in performance and torque

Not lacking in space: The Discovery cabin

Freelander from 1998

The first Land Rover equipped with independent suspension and featuring monocoque construction. Available in two versions: a closed five-door Station Wagon and a three-door available either with a hardback or a softtop. Engines are from Rover Cars.

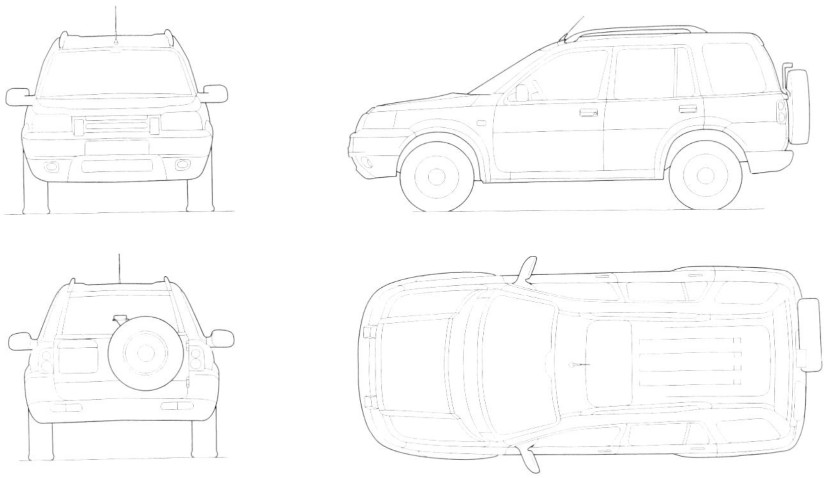

Freelander 1.8	Freelander 2.0 TD
1998 Model Year	1998 Model Year
Water-cooled four cylinder engine with electronic fuel injection	Water-cooled four cylinder turbo-diesel engine with direct injection
1796cc	1994cc
80.0mm x 89.3mm	84.5mm x 88.9mm
10.5:1	19.5:1
120bhp(88kw)/5500rpm	97bhp(72kw)/4200rpm
165Nm/3250rpm	210Nm/2000rpm
Five-speed gearbox; permanent four-wheel-drive with a viscous-coupling and electronic traction control	Five-speed gearbox; permanent four-wheel-drive with a viscous-coupling and electronic traction control
3.25:1, 1.89:1,1.22:1, 0.85:1, 0.65:1, R 3.00:1	3.25:1, 1.89:1,1.22:1, 0.85:1, 0.65:1, R 3.00:1
195R15 or 216R16	195R15 or 216R16
Front independent suspension with MacPherson strut lower arms, coil springs and anti-roll bar; rear independent suspension with MacPherson strut lower arms, coil springs and anti-roll bar	Front independent suspension with MacPherson strut lower arms, coil springs and anti-roll bar; rear independent suspension with MacPherson strut lower arms, coil springs and anti-roll bar
Servo-assisted brakes (discs front, drums at rear), ABS	Servo-assisted brakes (discs front, drums at rear), ABS
Rack and pinion steering with power assistance	Rack and pinion steering with power assistance
11.6m	11.6m
2.55m	2.55m
1.53/1.54m	1.53/1.54m
4.39m	4.39m
1.77m	1.77m
1.71m	1.71m
102mph	102mph
1425kg	1525kg
400kg	400kg

The Freelander utilises the MGF's four cylinder engine

Freelander drive train with viscous-coupling

	Forward Control 1962 until 1973	**Lightweight 1968**	**101 1975**
	Land Rover's first forward control model produced between 1962 and 1973. Available with petrol and diesel engines.	Built between 1968 and 1984, the Half Tonne Land Rover or "Lightweight" was built exclusively for the military and was air-portable.	Another exclusive military contract. This forward control shared some components with the Range Rover. It was built between 1975 and 1978.
	109 FC Series IIA 1962-1966 Series IIB 1967-1975 Model Year 1966	**Lightweight** Model Year 1971	**101** Model Year 1975
Engine:	Water-cooled four cylinder engine, Solex carburettor	Water-cooled four cylinder engine, Zenith carburettor	Water-cooled V8 cylinder engine, two Zenith Stromberg carburettors
Cylinder Capacity:	2286cc	2286cc	3528cc
Bore x Stroke:	90.5mm x 88.9mm	90.5mm x 77.0mm	88.9mm x 71.1mm
Compression:	7.0:1	7.0:1	8.5:1
Power output:	77bhp/4250rpm	77bhp(149kw)/4250rpm	135bhp/4750rpm
Torque:	168Nm/2500rpm	168Nm/2500rpm	278Nm/4000rpm
Power train:	Four-speed gearbox with reverse and transfer box; third and top gear synchromeshed; selectable four-wheel- or two-wheel-drive (rear) in high range; permanent four-wheel drive in low range	Four-speed gearbox with reverse and transfer box, third and top gear synchromeshed; selectable four-wheel- or two-wheel-drive (rear) in high range; permanent four-wheel-drive in low range	Four-speed gearbox with reverse and transfer box; permanent four-wheel-drive, lockable centre differential
Gears:	3.00:1, 2.04:1, 1.38:1, 1:1, R 2.54:1; transfer box giving 1.53:1 (high) 2.92:1 (low)	3.00:1, 2.04:1,1.38:1, 1:1, R 2.54:1; transfer box giving 1.148:1 (high) 2.891:1 (low)	4.05:1, 2.41:1, 1.61:1, 1:1, R 4.22:1; transfer box giving 1.17:1, (high) and 3.32:1 (low)
Wheels:	9.00 x 16	7.50 x 16	9.00 x 16
Suspension:	Live front and rear axles, semi-elliptic leaf springs, telescopic shock absorbers	Live front and rear axles, semi-elliptic leaf springs, telescopic shock absorbers	Live front and rear axles, semi-elliptic leaf springs, hydraulic, telescopic dampers
Brakes:	Hydraulic brakes (drums)	Hydraulic brakes (drums)	Hydraulic servo-assisted drum brakes
Steering:	Recirculating ball, lower-geared steering	Worm and nut with recirculating ball steering	Worm and nut with recirculating ball steering with power assistance
Turning circle:	13.7m	11.6m	11.3m
General technical specification:			
Wheelbase:	2.77m	2.24m	2.57m
Track:	1.36m	1.31m	1.52/1.55m
Length:	4.90m	3.65m	4.13m
Width:	1.92m	1.52m	1.78m
Height:	2.59m	1.95m	2.23m
Topspeed:	60mph	65mph	85mph
Unladen weight:	1902kg	1454kg (stripped down: 1205kg)	2656kg
Max. payload:	1531kg	561kg	1000kg

Production figures

Year	Total	Land Rover/ Defender	Range Rover	Discovery
1947-48	•	80	•	•
1948-49	•	8,000	•	•
1949-50	•	16,085	•	•
1950-51	•	17,360	•	•
1951-52	•	19,591	•	•
1952-53	•	18,570	•	•
1953-54	•	20,135	•	•
1954-55	•	28,882	•	•
1955-56	•	28,365	•	•
1956-57	•	25,775	•	•
1957-58	•	28,656	•	•
1958-59	•	28,371	•	•
1959-60	•	34,168	•	•
1960-61	•	35,148	•	•
1961-62	•	37,139	•	•
1962-63	•	34,304	•	•
1963-64	•	42,569	•	•
1964-65	•	45,790	•	•
1965-66	•	47,941	•	•
1966-67	•	44,191	•	•
1967-68	•	44,928	•	•
1968-69	•	50,561	•	•
1969-70	47,624	47,538	86	•
1970-71	59,200	56,663	2,537	•
1971-72	57,955	52,445	5,510	•
1972-73	56,243	49,724	6,519	•
1973-74	53,773	45,169	8,604	•
1974-75	63,197	52,681	10,516	•
1975-76	60,721	48,514	12,207	•
1977	51,119	41,452	9,667	•
1978	57,412	46,172	11,240	•
1979	54,309	42,936	11,373	•
1980	60,906	51,198	9,708	•
1981	51,500	41,059	10,441	•
1982	52,181	38,926	13,255	•
1983	40,768	28,586	12,182	•
1984	37,447	25,562	11,885	•
1985	37,086	23,772	13,314	•
1986	33,689	19,195	14,494	•
1987	41,290	20,475	20,815	•
1988	46,250	22,229	24,021	•
1989	55,000	22,738	28,513	3,749
1990	68,621	21,363	24,202	23,056
1991	54,574	18,906	16,408	19,260
1992	57,332	17,651	15,334	24,347
1993	68,159	17,590	14,949	35,620
1994	94,472	22,224	17,824	54,424
1995	127,287	27,670	29,619	69,998
1996	126,334	31,856	29,441	65,037
1997	127,962	36,169	30,066	58,352
total:	2,399,020	1,627,072	414,730	353,843

Production figures breakdown by type

Model	Year	Total
Land Rover Series I:	1948-1958	211,499
Land Rover Series II/IIA:	1958-1971	549,311
Land Rover Series III:	1971-1983	538,862
Land Rover 90/110/Defender:	1983-1997	327,400
Range Rover Series I:	1970-1996	317,615
Range Rover Series II:	1994-1997	97,115
Discovery:	1989-1997	353,843
Freelander:	1997	3,375

Note

From 1977, the Rover production year matched up with the calendar year for the first time. Numbers of limited edition vehicles are incorporated into the totals. The same goes for Minerva and Tempo vehicles. Santana produced 287,067 vehicles

List of Sources

TEXT

Chris Bennett
Land Rover
Osprey Publishing Ltd., London 1992

Robert Brandes
Wegbereiter/Land Rover Kaufberatung
Oldtimer Markt 9/1995
VF Verlag, Mainz

Enzyklopaedie des Automobils
Weltbild Verlag GmbH
Augsburg 1989

Olaf v. Fersen
Ein Jahrhundert Automobiltechnik
VDI-Verlag, Duesseldorf 1986

G.N. Georgano
The Complete Encyclopedia of Commercial Vehicles
Krause Publications, Iola Wisconsin 1979

Mike Hallett
The Land Rover Story
Off-Road and Four-Wheel-Drive, Croydon 1988

The Land Rover Directory
Redwood Burn Ltd., Trowbridge 1986

Volker C. Manz
Seltsame Land Rover aus Andalusien
Gelaendewagen Magazin 2/1997

A Motoring Heritage
John Bartholomew & Son Ltd.
Edinburgh 1976

Matthias Pfannmueller
Mit Tempo durch die Zeit
autovision, Hamburg 1996

Lindsay Porter
Land Rover Series I, II and III
Haynes Publishing Group, Somerset 1992

Graham Robson
The Land Rover-Workhorse of the World
David and Charles Publishers Ltd.
Devon 1976

Graham Robson
Die Rover Story
Econ Verlag GmbH 1978

Graham Robson
The Range Rover/ Land Rover
Redwood Burn Ltd., Trowbridge 1979

Halwart Schrader
Land Rover 1948 – 1955
Schrader Verlag 1995

Tom Sheppard
The Land Rover Experience
Land Rover by Butler & Tanner Ltd.
Frome and London 1993

Ken and Julie Slavin/George Mackie
Land Rover – The Unbeatable 4x4
Haynes Publications Inc., Somerset 1989

James Taylor
The Land Rover 1948-1988
Motor Racing Publications Ltd.
Croydon 1988

James Taylor
Land Rover Discovery
The Enthusiast's Companion
Motor Racing Publications Ltd.
Croydon 1994

James Taylor/Nick Dimbleby
Range Rover – The Complete Story
Crowood Press Ltd., Marlborough 1995

John Tipler
Land Rover – The Original 4x4
Crowood Press Ltd., Marlborough 1995

PHOTOS

Auto Union Archiv, Ingolstadt	9
Tom Barton, Birmingham	15, 25, 39, 43, 55, 61, 64, 65, 66, 72, 75, 77, 92, 93
British Motor Industry Heritage Trust, Gaydon	12, 14, 16, 20, 21, 22, 23, 24, 26, 27, 28, 29, 30, 31, 32, 33, 34, 36, 37, 38, 39, 40, 41, 42, 44, 45, 46, 47, 49, 50, 51, 52, 53, 54, 56, 57, 59, 60, 62, 63, 65, 66, 67, 69, 70, 71, 72, 73, 74, 78, 82, 86, 87, 88, 89, 90, 91, 92, 101
Bert Buurman, The Netherlands Photo Archive	27
Daimler-Benz Classic Archiv, Stuttgart	9
Daimler-Benz AG Werkarchiv Hamburg	9, 10, 30, 44, 47, 48, 49, 52, 79, 80, 81
Andrew Entwistle, Croydon	144
Dr. Dieter Günther, Hamburg	58, 68, 75, 83
Torsten Herborn, Wiesbaden	144
Klaus Huber-Abendroth, AUTO, Berlin	110, 111
Massey Ferguson Ltd., Coventry	17
Archiv Neubauer, Hamburg	8, 10, 11, 68
Matthias Pfannmüller, Hamburg	1, 5, 6, 20, 51, 52, 61, 65, 106, 109, 110, 111, 114, 115, 118, 119, 120, 121, 123, 126, 127, 132, 144
Santana Motor S.A., Linares	84, 85
Boris Schmidt, Frankfurt	10, 11, 78, 83, 85
Michael Specht, Hamburg	108
Rover Deutschland GmbH, Neuss	54, 71, 74, 75, 94, 96, 100, 105, 107, 108, 109, 135, 136
Rover Group Ltd., Warwick	92, 94, 95, 97, 98, 99, 102, 103, 104, 112, 113, 116, 117, 120, 122, 138, 139
WBI, Köln	126, 127
Steve Wilks, East Grinstead	4, 5, 13, 18, 19

Acknowledgements

This book was written with the help of Rover Germany Ltd. and the following:

- Tom Barton, Birmingham
- Hubert Bergmann, Terry Donovan and Mike Gould, Rover Group Ltd., Warwick
- Felix Bittmann, Hamburg
- Hellmuht Carroux, Hamburg
- Annemarie Casey, Galway
- Ursula Deleonibus, Eppstein
- Kirsten Diedrichsen British Consulate, Hamburg
- Lothar Franz, Ingolstadt
- Joerg Friese, WBI, Cologne
- Gerd Frommherz, Rheinselden
- Torsten Herborn, Wiesbaden
- Georges H. Keller, Rolls Royce Motor Cars International SA, Au Glapin
- Spen King, Leamington Spa
- Hans Joachim Klersy, Mainz
- Norman Krandall, Eyreford
- Ulli Kraemer-Rothhaar, Mainz
- Ingrid Langeveld, Valerie Hanke, Ruediger Assion, Markus Hauf and Dieter Laxy, Rover Deutschland GmbH, Neuss
- Dr. Gerold Lingnau, Eppstein
- Joern Lissow, Hamburg
- Moses Ludel, Reno
- Dietrich Matthes, Darmstadt
- Hans Otto Neubauer, Hamburg
- Evan Owen, Brynsiencyn/Anglesey
- Karam Ram, BMIHT Gaydon
- Reimer Schulz, Hamburg
- Olaf Schwientek, Selters/Taunus
- John Smith, Fairseat
- William Thomas, Wiesbaden
- Dietrich Vahldieck, Daimler-Benz AG, Hamburg
- Bernhard Weinbacher, Muenchen
- Christian Wiegenhagen, Ottersberg
- Anna and John Wilkins, Brill
- Patrick Wilks, Birmingham
- Steve Wilks, East Grinstead
- Walter Wolf, Riedstadt
- Jose A. Navarro Yague, Santana-Motor S.A. Linares
- Volker Zander, Lueneburg

The Authors

Matthias Pfannmueller was caught by the car bug very early on in life and was fortunate enough to transform his hobby into a career. Now a freelance journalist living in Hamburg, Matthias writes and photographs for numerous European magazines and newspapers. Bowled over by the subject matter and just in time for Land Rover's 50th Anniversary, he bought an 80inch Series I

Boris Schmidt studied Politics, Law and Journalism at the University of Mainz.
For eight years, the motor industry has been his bread and butter. As a journalist on Germany's leading quality daily, the Frankfurter Allgemeine Zeitung, he writes for the "Technology and Motors" section.
He has already owned two Land Rovers and is currently searching for the third

Translator Paul Entwistle studied German at the University of Manchester before moving to Berlin where he worked as a freelance journalist.
After joining Land Rover World magazine in 1994, he finally achieved a lifetime ambition and acquired his first Lightweight. Paul is currently employed by Rover Group Limited as a press officer.
He is already looking for his next Lightweight